Advances in Biology

This is a numbered series focused on the latest research in biology.

nova
science publishers
www.novapublishers.com

Advances in Biology

Advances in Biology. Volume 1
Charles D. Grant (Editor)
2022. ISBN: 979-8-88697-201-6 (Hardcover)
2022. ISBN: 979-8-88697-218-4 (e-book)

Charles D. Grant
Editor

Advances in Biology

Volume 1

nova
science publishers
www.novapublishers.com

Library of Congress Cataloging-in-Publication Data

ISBN: 979-8-88697-201-6

Published by Nova Science Publishers, Inc. † New York

Contents

Preface vii

Chapter 1 **Enzymes for the Synthesis of Rare Sugars** 1
Can Li, Mesfin Angaw Tesfay, Xin Wen, Yuhang Ning, Huibin Lin, Chengjia Zhang, Yilin Ren, Jianqiang Lin and Jianqun Lin

Chapter 2 **Functional and Bioenergetic Effects of ATP-Sensitive Potassium Transport in Mitochondria** 49
Olga V. Akopova

Chapter 3 **Bilirubin Interactions with Different Proteins and Implications of These Interactions** 85
Nikola Gligorijević and Simeon Minić

Chapter 4 **Overview of Panton-Valentine Leukocidin-Expressing *Staphylococcus Aureus* Pneumonia and Its Therapeutic Strategy** 123
Jin-Wei Liu and Wen-Liang Yu

Chapter 5 **Biochemical Thermodynamics** 153
E. Méndez

Chapter 6 ***In-Vitro*** **Activity of Ceftolozane/Tazobactam against *Pseudomonas Aeruginosa* and *Enterobacterales* Isolates from Taiwan** 185
Yee-Huang Ku, Po-An Su, Yin-Ching Chuang and Wen-Liang Yu

Chapter 7 **Elevated Levels of Caspase-3 in Cerebrospinal Fluid of Newborns with Perinatal Asphyxia: A Biomarker of Failing Mitochondrial Bioenergetics** 205
Puneet K. Samaiya, Ashok Kumar and Sairam Krishnamurthy

Index 215

Preface

This volume includes seven chapters that describe recent advances in biology. Chapter One reviews and discusses the enzymes and methods for bioproduction of the rare sugars of D-allulose, D-allose, allitol, L-xylulose, L-xylose, and L-ribose. Chapter Two analyzes the properties of mKATP channels and functional and bioenergetic effects of ATP-sensitive K+ transport. Chapter Three describes interactions of bilirubin with several proteins and the effects of these interactions on the protein's structure and function. Chapter Four provides an overview of Panton Valentine Leukocidin-producing *Staphylococcus Aureus* pneumonia and its therapeutic strategy. Chapter Five discusses the reunification of chemical and biochemical thermodynamics. Chapter Six includes a study of the in-vitro activity of eftolozane/tazobactam against pseudomonas aeruginosa and enterobacterales isolates from Taiwan. Finally, Chapter Seven examines elevated levels of caspase-3 in cerebrospinal fluid of newborns with perinatal asphyxia as a biomarker of failing mitochondrial bioenergetics.

Chapter 1 - Rare sugars are defined as monosaccharides and their derivatives that are rare in nature by the International Society of Rare Sugars. Most rare sugars taste sweet but are not metabolized by humans so they can be used as low-calorie sweeteners. Some rare sugars have health care or medicinal functions like blood glucose suppressing or antiviral and anticancer functions, etc. So, rare sugars have many potential uses in the food and medicine industries. For large-scale production of rare sugars, isolation from natural sources is not practical. Chemical synthetic methods are sophisticated and can produce toxic byproducts. Biotechnological production is superior to the above methods. In this chapter, the enzymes and methods for bioproduction of the rare sugars of D-allulose, D-allose, allitol, L-xylulose, L-xylose, and L-ribose will be reviewed and discussed.

Chapter 2 - ATP-sensitive potassium uptake conducted by mitochondrial ATP-sensitive potassium channels (mKATP channels) is presently recognized as a powerful tool of tissues protection under several pathophysiological states and conditions. Cytoprotective signaling triggered by mKATP channels opening is largely based on the modulation of mitochondrial bioenergetics. ATP-sensitive K+ transport is an all-round modulator of mitochondrial functions and energy-dependent processes: potassium cycle, oxygen consumption, ATP synthesis, Ca2+ uptake and ROS production. The modulation of these processes is based on the abundance of mKATP channels in mitochondrial membrane, on the rate of potassium transport, and the contribution of ATP-sensitive K+ transport to oxygen consumption in mitochondria. For understanding the impact of mKATP channels opening on mitochondrial bioenergetics, it is important to know how functional effects of ATP-sensitive K+ transport (the rate of K+ uptake and its share in oxygen consumption) are transformed into bioenergetic ones: the modulation of membrane potential, uncoupling of the respiratory chain and related energy-dependent processes (Ca2+ transport, ATP synthesis). It is strangely enough, but there is still no consensus in the literature about the direct effects of ATP-sensitive K+ transport on mitochondrial functions. The use of pharmacological modulators is the most common way to assess mKATP channels functioning in mitochondria. However, one crucial point of these studies is the lack of specificity of commonly used pharmacological modulators of mKATP channels, the openers (diazoxide, pinacidil, nicorandil, chromacalim), and the blockers (glibenclamide and 5-hydroxydecanoate). In this chapter the properties of mKATP channels, functional and bioenergetic effects of ATP-sensitive K+ transport are analyzed based on the authors' studies on isolated mitochondria. Also, the literary data on the properties of mKATP channel, cellular mechanisms of tissue protection, bioenergetic effects of mKATP channels opening, and the prospectives in mKATP channels studies are discussed.

Chapter 3 - Bilirubin is a tetrapyrrole, yellow pigment, and it is a degradation product created in heme metabolism. This molecule is predominantly present in its unconjugated form in circulation, and it is mainly bound to human serum albumin. The presence of small amounts of free, unconjugated bilirubin enables its interactions with other circulation components. In conditions like Gilbert syndrome or diabetes, the concentration of unconjugated bilirubin increases in the blood, making other interacting partners from circulation more important for its metabolism. In the liver, bilirubin is converted to conjugated form, which is then excreted from the organism. While the significant increase of bilirubin is toxic, research suggests that small increases may be beneficial since bilirubin has antioxidative and anticancer potential. In this chapter, interactions of bilirubin with several proteins will be described together with the effects of these interactions on the protein's structure and function.

Chapter 4 - Panton Valentine leukocidin (PVL)-producing *Staphylococcus aureus* pneumonia is not a common but extremely life-threatening disease. Without immediate recognition and appropriate management, it can result in at least 75% mortality rate higher than PVL-negative strains. PVL is often found in community-associated methicillin-resistant *S. aureus* as well as community-associated methicillin-sensitive *S. aureus* isolates. The risk host factors of PVL-positive *S. aureus* pneumonia may include an original health individual, especially children and adolescents with a preceding respiratory virus infection, and the people close skin-to-skin contact with each other, like athletes, military trainees, jail inmates, etc. The pathogenesis of PVL may be attributed to the genes lukS-PV and lukF-PV, a pore-forming cytotoxin that can directly and indirectly cause polymorphonuclear leukocyte (PMN) destruction, resulting in severe leukopenia. Consistent with the PMN cytolytic activity of PVL, neutropenia is a frequent clinical finding associated with PVL-positive *S. aureus* necrotizing pneumonia. In addition, chest radiography may picture an initial mild pneumonia patch and can rapidly progress to diffuse multilobar consolidation with cavity formation around both lung fields. Hence, patients can present with acute respiratory distress syndrome and septic shock-like syndrome that require admission to an intensive care unit upon arrival of the emergent department or few days after admission. Currently, there is still no consensus for the management of this disease. Nevertheless, the detrimental effects of the infection are mainly caused by the production of PVL rather than the microorganism itself, therefore, an effective treatment should not only target at the eradication of *S. aureus* but additionally at eliminating the virulence of the toxin. The authors review the therapeutic strategies up to 2022 and recommend the following two methods. One is the combination of β-lactam antibiotics or glycopeptides with one kind of bacteriostatic antibiotics, acting as protein synthesis inhibitors, like clindamycin, linezolid, and tigecycline. The other is the addition of a combination of intravenous linezolid with intravenous clindamycin and rifampicin to the initial empirical antibiotic regimens if strongly suspected PVL-positive *S. aureus* pneumonia. Meanwhile, adjunctive therapy like intravenous immunoglobulin (IVIG) is considered to inactivate the toxin as soon as possible. IVIG is believed to neutralize PVL pore formation and the cytopathic effects of PVL, bringing beneficial benefits to the prognosis. For those critical patients with unstable vital signs, the invasive life-supportive methods like mechanical ventilation or even the use of extracorporeal membrane oxygenation should be initiated without delay. Moreover, the cytokine adsorption therapy has emerged as a valuable adjuvant therapy and it can rapidly reduce the inflammatory status, thus lowering the need or dosage of vasopressor agents. In conclusion, due to the considerable lethality of PVL-expression *S. aureus* pneumonia, early identification of the PVL gene is so crucial that physicians can take the direct therapeutic strategies to enhance the success of saving patient's life and perhaps could avoid further invasive life-sustained treatment.

Chapter 5 - Since the studies pioneered by Hans Krebs in the '50s, the thermodynamic basis of enzyme-catalyzed reactions has received considerable attention. After the implementation of the biochemical reference state, the prediction power of thermodynamic calculations has increased notably. However, there is still a gap in the amount of information that prevents the wide application of biochemical thermodynamics to different processes. Group-contribution method, quantum chemical calculations, and big data science emerge as a solution to this problem. In this chapter, biochemical thermodynamics is considered after a brief historical revision, some basic thermodynamic foundations, the introduction of the different reference states, as well as the curated source of thermodynamic and experimental data that allow for the precise calculation of missing data. Novel calculation procedures will be considered. Finally, the reunification of chemical and biochemical thermodynamics will be discussed.

Chapter 6 - *Background.* Ceftolozane/tazobactam (TOL/TAZ), a new antipseudomonal cephalosporin combined with a β-lactamase inhibitor, is indicated for the treatment of complicated intra-abdominal infections (requiring in combination with metronidazole), complicated urinary tract infections, and nosocomial or ventilator-associated pneumonia.

Methods. Activities of TOL/TAZ and ten comparators (cefepime, cefoperazone/sulbactam, ceftazidime, ciprofloxacin, colistin, imipenem, levofloxacin, meropenem, piperacillin/tazobactam [PIP/TAZ] and tigecycline) were evaluated against isolates collected in Taiwan. The susceptibility was tested by broth microdilution methods.

Results. A total of 599 isolates were collected during 2012-2015, including intra-abdomen (n = 223), respiratory tract (n = 153), urine (n = 149), skin and soft tissue (n = 55), blood (n = 11) and others (n = 8). TOL/TAZ demonstrated a good activity against Enterobacterales isolates (n = 446; 84.8% susceptibility; susceptible breakpoint, ≤ 2 μg/mL). TOL/TAZ was highly active versus Citrobacter koseri (n = 15), Klebsiella oxytoca (n = 30), Morganella morganii (n = 15), Proteus mirabilis (n = 30), and Escherichia coli (n = 125), with MIC90 values of ≤ 1 μg/mL. TOL/TAZ had the best overall activity versus P. aeruginosa (94% susceptibility; MIC90 = 4 μg/mL; susceptible breakpoint, ≤4 μg/mL), in comparison to imipenem (59% susceptibility, MIC90 >8 μg/mL), meropenem (71% susceptibility, MIC90 >8 μg/mL), and colistin (100% intermediate, MIC90 = 2 μg/mL). The susceptibility rates of TOL/TAZ for K. pneumoniae isolates susceptible and non-susceptible to ceftazidime were 75.2% and 31.1%, respectively.

Conclusion. TOL/TAZ demonstrated potent in-vitro activity against P. aeruginosa and Enterobacterales isolates in Taiwan. TOL/TAZ is a good option for treatment of infections caused by these organisms.

Chapter 7 - Mitochondria act as metabolic hubs within the cell, which are essential for energy production to fulfill the bioenergetic and biosynthetic demands of the developing brain. Failing mitochondrial bioenergetics can play a central role in neuronal cell death in pathological conditions like perinatal asphyxia (PA), which can be associated with insult progression over days and the development of a more severe condition hypoxic-ischemic encephalopathy (HIE) in the neonates. Therefore, to evaluate the role of mitochondrial bioenergetics in PA/HIE, the authors estimated apoptotic markers like caspase-3 in cerebrospinal fluid (CSF) of newborns with asphyxia (n = 42) and control (n =17) at birth. Apgar scores < 7 at 5 min were considered as asphyxic. Further, through sarnat scoring, the asphyxic neonates were divided into different stages of HIE. CSF sample was collected as sample-1 on day-1 and sample-2 on day-7 after asphyxia. The authors observed a significant ($p < 0.05$) increase in the levels of caspase-3 in the HIE stage (s)-III compared to s-II and s-I and a marked increase in the control group in sample-1. Further, the levels of caspase-3 were markedly elevated in HIE s-III compared to s-II and s-I in sample-2, which depicts the persistence of neuronal insult. The increased levels of caspase-3 in the CSF of neonates with different stages of HIE after asphyxia indicate activation of apoptosis in the developing brain of neonates, which can lead to neurological abnormalities in the later stage of life.

Chapter 1

Enzymes for the Synthesis of Rare Sugars

Can Li[1], Mesfin Angaw Tesfay[2], Xin Wen[2], Yuhang Ning[2], Huibin Lin[3], Chengjia Zhang[2], Yilin Ren[4], Jianqiang Lin[2,*] and Jianqun Lin[2,†]

[1]School of Biological Engineering, Qilu University of Technology, Jinan, China
[2]State Key Laboratory of Microbial Technology, Shandong University, Qingdao, China
[3]Shandong Academy of Chinese Medicine, Jinan, China
[4]Qingdao Longding Biotech Limited Company, Qingdao, China

Abstract

Rare sugars are defined as monosaccharides and their derivatives that are rare in nature by the International Society of Rare Sugars. Most rare sugars taste sweet but are not metabolized by humans so they can be used as low-calorie sweeteners. Some rare sugars have health care or medicinal functions like blood glucose suppressing or antiviral and anticancer functions, etc. So, rare sugars have many potential uses in the food and medicine industries. For large-scale production of rare sugars, isolation from natural sources is not practical. Chemical synthetic methods are sophisticated and can produce toxic byproducts. Biotechnological production is superior to the above methods. In this chapter, the enzymes and methods for bioproduction of the rare sugars of D-allulose, D-allose, allitol, L-xylulose, L-xylose, and L-ribose will be reviewed and discussed.

Keywords: rare sugar, enzyme, catalysis, biotransformation

---

* Corresponding Author's Email: jianqianglin@sdu.edu.cn.
† Corresponding Author's Email: jianqunlin@sdu.edu.cn.

In: Advances in Biology. Volume 1
Editor: Charles D. Grant
ISBN: 979-8-88697-201-6

Introduction

Sugars are not only a source of energy but also mediate a variety of recognition processes that are central to human health. As a result, carbohydrates can be used as prebiotics in the food and pharmaceutical industries to stimulate the immune system, as low-calorie sweeteners to control diabetes, and as starting materials of L-nucleosides for the production of anticancer and antiviral drugs. However, only a small number of all possible monosaccharides are found in nature in sufficient amounts to allow their commercial exploitation (Beerens et al., 2012). Sugars that occur naturally in small amounts are called rare sugars.

Rare sugars are sugars that occur naturally in extremely small amounts, usually because they are products of intermediary cellular metabolism and are converted within the cell to another form as soon as they are created. According to the International Society of Rare Sugars (ISRS) (Granström et al., 2004), established in Japan, rare sugars are defined as monosaccharides and their derivatives that are rare in nature.

Nowadays, rare sugars are usually biotransformed starting from cheap and widely available substrates. In this chapter, the enzymes for bioproduction of D-allulose, D-allose, allitol, L-xylulose, L-xylose, and L-ribose are introduced, and the biotransformation methods of the above rare sugars are discussed. In addition, the research progress of the above rare sugar biotransformation was reviewed.

D-Tagatose 3-Epimerase (DTE) for D-Allulose Bioproduction

Backgrounds

D-Allulose (or D-psicose), also known as D-ribose-2-hexulose, is the C-3 epimer of D-fructose (Figure 1). The molecular formula of D-allulose is $C_6H_{12}O_6$ and it has a molecular weight and melting point of 180.16 g/mol, and 96°C respectively. D-Allulose is very soluble in water, and 291 g of D-allulose can be dissolved in 100 g of water at 25°C. D-Allulose is a white crystalline substance, odorless, and not easy to absorb moisture. It has a consistent crystal structure similar to β-D-pyranose as determined by X-ray crystal structure analysis.

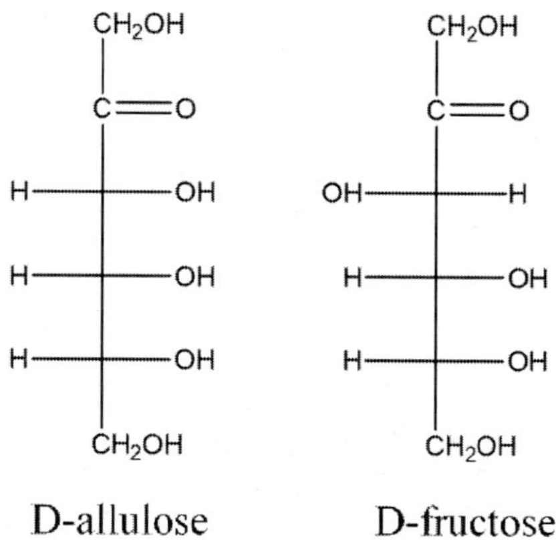

Figure 1. Chemical structure of D-allulose and D-fructose.

D-Allulose has a sweet taste and can be used as an "energy-free" sweetener because it has 70% sweetness and only 1% energy compared to sucrose. D-Allulose is regarded as GRAS (Generally Recognized as Safe) and allowed to be used as food and dietary supplements by the U.S. Food and Drug Administration (FDA) in 2011 and in some other countries. The study of the metabolic characteristics of D-allulose in rats shows that it is mainly absorbed in the small intestine and enters the blood circulation after the oral administration of D-allulose. About 11-25% of D-allulose is excreted through urine after 24 hours. The unabsorbed D-allulose enters the large intestine and generates short-chain fatty acids through the fermentation of intestinal microbiota. Finally, about 8-13% of D-allulose that is not metabolized is excreted through feces. The metabolism of D-allulose in the human body is different from that in rats. The D-allulose that is absorbed in the small intestine is not metabolized to produce energy, and will not affect blood glucose concentration or insulin level. In addition, a small part of D-allulose reaching the large intestine cannot be decomposed by microorganisms in the intestine. D-Allulose has a variety of physiological functions: it has very low energy (0.007 kcal/g) and can prevent obesity by reducing the activity of fat synthase; it can be used as adjunctive therapy for type II diabetes by lowering blood sugar; it has antioxidant activity, strong scavenging ability of reactive oxygen species (ROS) and reducing the ability of glutathione; it has auxiliary nerve therapy function and neuroprotective function. So, D-allulose is a functional and low-energy sweetener.

Source and Properties of DTEase Family Enzymes

According to the Izumoring strategy for the synthesis of rare sugar (Itoh et al., 1995; Yoshida et al., 2007; Chan et al., 2012; He et al., 2016), D-allulose can be produced by oxidation of allitol using oxidoreductase or by isomerization of D-fructose using isomerase. Various enzymes such as D-tagatose 3-epimerase, aldose isomerase, polyol dehydrogenase, oxidoreductase, etc. can be used in D-allulose biotransformation. Allitol is expensive and difficult to obtain hence, the D-fructose isomerization method has attracted more attention and is used in industrial applications to produce D-allulose.

There are mainly two kinds of enzymes that isomerize D-fructose to produce D-allulose: D-tagatose 3-epimerase (DTE, EC 5.1.3.31) and D-allulose 3-epimerase (DPE, EC 5.1.3.30), collectively referred to as DTEase family enzymes. Izumori et al. found D-tagatose 3-epimerase (named because its optimal substrate is D-tagatose) in *Pseudomonas cichorii* ST-24 and realized the transformation from D-fructose to D-allulose for the first time (Izumori et al., 1993). Kim et al. reported an enzyme similar to DTEase in *Agrobacterium tumefaciens*, which can efficiently catalyze the conversion between D-fructose and D-allulose, which is named D-allulose 3-epimerase because its optimum substrate is D-allulose (Kim et al., 2006).

Table 1. Comparison of enzymatic properties of DPEase from different strains

Conditions / Strains	Optimal Temperature (°C)	Optimal pH	Half-life at 50°C (min)	Substrate with highest specificity	Equilibrium ratio between D-allulose and D-fructose
A. tumefaciens	50	8.0	63.5	D-allulose	32:68 (30°C)
C. scindens	60	7.5	108	D-allulose	28:72 (50°C)
C. cellulolyticum	55	8.0	>120	D-allulose	32:68 (60°C)
Ruminococcus sp.	60	7.5-8.0	>240	D-allulose	28:72 (60°C)

The coding genes of the DTEase enzyme have been identified from *Clostridium cellulolyticum* H10, *Ruminococcus sp.*, *Clostridium scindens* (Zhang et al., 2013), and *Desmospora sp.* It is found that the family enzymes from different microbial sources have different enzymatic properties. The comparison of enzymatic properties of DPEase from *A. tumefaciens*, *C. scindensa*, *C. cellulolyticu,* and *Ruminococcus sp.* are shown in Table 1.

Although the amino acid sequence homology of family enzymes is not high (23%~92%), all family enzymes can catalyze the mutual transformation of fructose and D-allulose.

Crystal Structure of DTEase

Among the DTEase family enzymes, DPEase from *A. tumefaciens* and *C. cellulolyticum* showed the closest tetramer form. In the tetramer of DPEase, the amino acid residues form 34 hydrogen bonds between 2 subunits, and the high catalytic activity of DPEase is attributed to the broad interfacial solvent accessible region, which is caused by the extensive interaction between the two dimers in the DTEase family enzymes. In addition, the DPEase expressed by the recombinant strain still showed excellent enzymatic properties. Heterologous expression of DPEase from *C. cellulolyticum* H10 has a longer half-life, higher kinetic parameters, and higher thermal stability.

Chan et al. analyzed the structure of the DPE enzyme derived from *Clostridium cellulolyticum* H10 (Chan et al., 2012), which is a tetrameric structure composed of four identical subunits. The structure of each subunit is a typical TIM barrel, consisting of eight repeated β-sheet and α-helix units constituting $(\beta/\alpha)_8$. A binding site for Mn^{2+} is shown in the center of each subunit structure. The combination of Mn^{2+} can increase the stability of the enzyme, thereby improving its activity. The key amino acid residues of its active center, metal ion binding site, and substrate binding site are highly conserved. The Mn ion forms octahedral coordination with two water molecules and four amino acids (Glu, Asp, His, and Glu), which are fully conserved among all ketose 3-epimerase enzymes. In addition, the three amino acid residues (Glu, His, and Arg) that interact with the O-1, O-2, and O-3 positions of D-fructose are also fully conserved.

Substrate Specificity of DTEase

Enzymes are special proteins with unique spatial structures. The active center of an enzyme molecule is selective to the stereochemical conformation of a substrate. When the conformation of the substrate matches the active center of the enzyme, the substrate can be well combined with the catalytic active center of the enzyme, so it has high enzyme activity. When the structure of a non-substrate molecule is similar to that of a substrate molecule, although it can still show certain activity, the specificity of an enzyme is not particularly good.

Using *bacillus* as the expression system, the effects of five D-ketoses with a similar structure to D-allulose and five D-aldoses with little structural similarity (D-allulose, D-fructose, D-tagatose, D-xylose, D-sorbose, D-allose, D-galactose, D-glucose, D-arabinose and D-mannose) on the activity of D-allulose 3-epimerase were studied. It was found that D-allulose 3-epimerase had certain substrate activity on the five D-type ketoses with a similar structure to D-allulose, and the sequence of substrate activity was D-fructose > D-tagatose > D-xylose > D-sorbose, while the isomerization activity of five D-aldoses with little structural similarity to D-allulose was very low or nonexistent. The study also found that although D-allulose 3-epimerase has certain substrate activity for most D-type ketoses, it does not affect the progress of the main reaction using a single substrate at a time. When the substrate of the DTEase enzyme is broad, it is necessary to study the influence of other substrates on its catalysis to screen the catalyst with better substrate specificity and higher conversion. These studies preliminarily analyzed the influence of the existence of other substrates on enzyme activity, and also laid a good theoretical and technical foundation.

Table 2. Substrate specificity of DPEase from *Ruminococcus sp.*

Substrate	Relative activity [a] (%)
D-Allulose	100 ± 1.1
D-Fructose	59.8 ± 0.6
D-Tagatose	24.2 ± 0.9
D-Xylulose	3.9 ± 0.2
D-Sorbose	2.9 ± 0.1
D-Allose	3.9 ± 0.3
D-Galactose	0.9 ± 0.2
D-Glucose	0.7 ± 0.1
D-Arabinose	0.5 ± 0.1
D-Mannose	N.D.[b]

[a] Activity relative to that for D-allulose.
[b] N.D.= not detected.

Catalytic Mechanism of DTEase

Ketose 3-epimerase has a unique catalytic mechanism, which is significantly different from other Epimerases. Its substrate does not need to be modified by phosphorylation. By analyzing the crystal structure of *A. tumefaciens* DPEase, it is speculated that its catalytic mechanism is that one Glu residue coordinates with Mn^{2+} to deprotonate the substrate at C-3 to form a cis-enediolate intermediate, and the other Glu residue immediately restores protonation of the intermediate in the opposite direction. Yoshida et al. proposed another catalytic mechanism by analyzing the crystal structure of *P. cichorii* DTEase, that is, in the process of substrate binding, G1u246 deprotonates the substrate C-3 to form a cis-enediolate intermediate with O2-C2-C3-O3 plane structure, and then protons are transferred from G1u246 to Glu152 through O-3. Finally, Glu152 promotes the protonation of C-3, and the two Glu return to the negative charge state (Yoshida et al., 2007). In recent years, although many DPE/DTE crystal structures have been successfully analyzed, there is still no clear certainty about the specific catalytic mechanism of DPE/DTE. The research on the catalytic mechanism of DTEase family enzymes is still in the preliminary stage, and the relationship between enzyme structure and catalytic function needs to be further studied. In addition, there are still some deficiencies in the thermal stability and substrate specificity of DTEase. In the production of rare sugars, the substrate specificity of DTEase family enzymes should be further used to realize the efficient and green production of functional rare sugars.

D-Allulose Biotransformation by Immobilized DTEase

Itoh et al. (1995) immobilized the DTEase on the bcw2503 carrier, and the immobilized enzyme was filled in the chromatographic column, which can be used continuously for 10 days. After the reaction, 90 g/L of D-allulose was obtained from 500 g/L of D-fructose (Itoh et al., 1995). Kim et al. (2006) immobilized the DPE enzyme from *Agrobacterium tumefaciens*, and produced 222 g/L D-allulose from 700 g/L D-fructose (Kim et al., 2006). Wei et al. (2013) immobilized the DPE enzyme of *Clostridium cellulolyticum H10* with chitosan, and 23% D-allulose was obtained from10 g/L D-fructose (Zhang et al., 2013). Tseng et al. (2014) fused the *dpe* from *Clostridium cellulolyticum* H10 with artificial lipophilic protein for enzyme immobilization, which improved the thermal stability of the enzyme (Tseng et al., 2014). Li et al. (2015) immobilized D-xylose isomerase from *Thermus thermophilus* and DPE enzyme from *Agrobacterium tumefaciens* on the spores of a *Saccharomyces cerevisiae* deficient strain △dit1 (the spore wall produced by this strain does not contain the outermost Di-tyrosine layer, and the outermost layer exposed is chitosan). 12% D-allulose was obtained from 3 g/L D-glucose (Li et al., 2015).

L-Rhamnose Isomerase for D-Allose Production

Backgrounds

D-Allose is the epimeric isomer of D-glucose at the C-3 position and the aldose isomer of D-allulose (Figure 2). Its molecular formula is $C_6H_{12}O_6$, with molecular weight and a melting point of 180.16 g/mol and 128°C respectively. It is soluble in water and insoluble in ethanol. D-Allose is an odorless white crystal that exists in some natural plant extracts and bacterial metabolites. It belongs to pentahydroxyaldehyde and can exist in linear and cyclic forms. The cyclic form is further divided into furan type and pyran type. In natural compounds, D-allose mostly exists in the form of β-pyran. Moreover, the acute and sub-chronic toxicity tests of D-allose show that D-allose is non-toxic.

D-Allose is an important rare sugar with good health care and medicinal value. It is reported to inhibit carcinogenesis and the proliferation of various cancer cell lines (Sui et al., 2005), have a positive effect on cancer radiotherapy and chemotherapy (Yamaguchi et al., 2008), inhibit high sodium-induced hypertension, have antioxidant effects, immunosuppressive and cell cryoprotectant effects (Ishihara et al., 2011; Hossain et al., 2004).

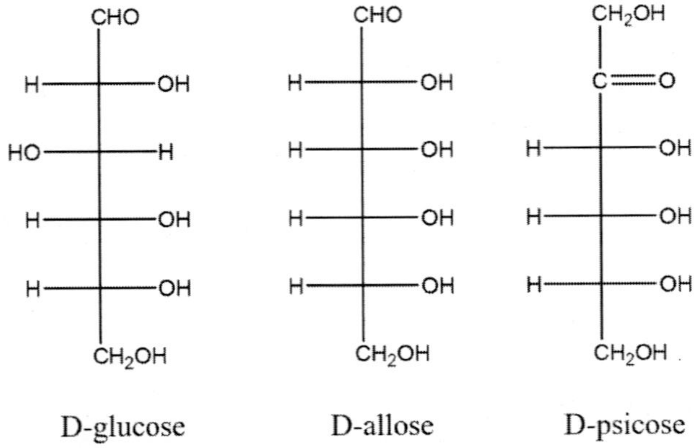

Figure 2. Chemical structure of D-glucose, D-allose and D-allulose.

In *E. coli*, D-allose can enter the cytoplasm through the transport system on the cell membrane, and participate in the glycolysis pathway. It is firstly combined with AlsB (D-allulose binding protein), and then transported into the cytoplasm by the action of AlsC (transport element) and AlsA (ABC transporter). In the cytoplasm, the conversion of D-allose to D-fructose-6-phosphate is achieved through the action of AlsK (D-Allokinase), AlsI (D-allose-6-phosphate isomerase), and AlsE (D-allulose-6-phosphate 3-epimerase). D-allose can also be utilized by *Aerobacter aerogenes*, but the metabolic mechanism of other bacteria has not been studied except *E. coli*.

The Source and Properties of L-Rhamnose Isomerase

Kenizumori (1997) found L-rhamnose isomerase (L-RhI) in *Pseudomonas stutzeri*, named because its optimal substrate is L-rhamnose (Bhuiyan et al., 1997). However, the L-RhI of *Pseudomonas stutzeri* is not suitable for the production of D-allose, because it forms a by-product of D-altrose (Figure 3), which makes it difficult to separate D-allose.

In addition, most L-RhI enzymes do not have the ability to catalyze the production of D-allose, such as the L-RhI from *E. coli*, *Bacillus halodurans* and *Thermotoga maritime*. Finding new L-RhI is the key to solving the biotransformation of D-allose. The L-Rhs derived from *P. stutzeri*, *C. saccharolyticus*, *B. subtilis* and *T. saccharolyticum* are found able to catalyze D-allose production, and their enzymatic properties are compared in Table 3.

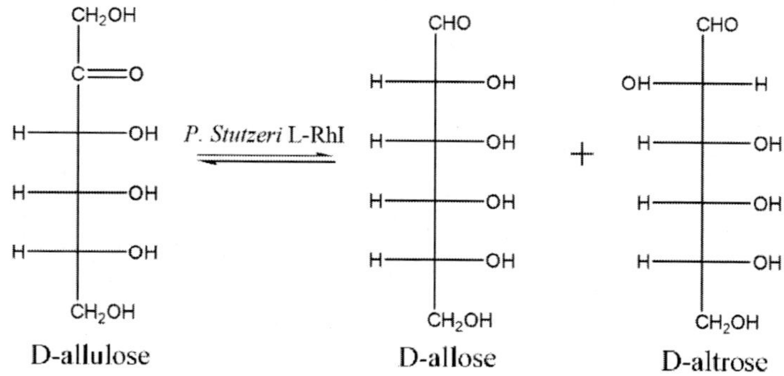

Figure 3. Chemical structure of products by *P. stutzeri* L-RhI from D-allulose.

Table 3. Comparison of enzymatic properties of L-RhIase derived from different bacteria

Conditions / Strains	Optimal Temperature(°C)	Optimal pH	Heat resistance (h)	With byproduct or not	Equilibrium ratio between D-allulose and D-allose
P. stutzeri	60	9.0	0.2 (50°C, 50%)	Yes	67:26 (50°C)
B. pallidus	65	7.0	1.0 (60°C, 50%)	No	65:35 (65°C)
C. saccharolyticus	90	7.0	6.0 (80°C, 90%)	No	67:33 (90°C)
B. subtilis	70	9.0	4.0 (50°C, 65%)	No	67:33 (70°C)
T. saccharolyticum	75	7.0	2.0 (70°C, 50%)	No	71:29 (70°C)

Table 4. Comparison of kinetic parameters of L-RI from various microbial sources

Microial sources of L-RhI	K_m/mmol.L^{-1}			K_{cat}/s^{-1}			$K_{cat}/K_m/Mm^{-1}s^{-1}$		
	LR[a]	DA[b]	DR[c]	LR	DA	DR	LR	DA	DR
P. stutzeri	11	42	38	171.67	5.02	15.77	15.61	0.12	0.415
B. pallidus Y25	4.89	41.8	34.9	68	34.5	14.18	13.91	0.83	0.41
T. saccharolyticum	3.53	121	148	179.32	33.9	21.5	50.8	0.28	0.15
T. maritima	37	NR	NR	146	NR	NR	3.94	NR	NR
M. loti	5	7.11	6.18	50.5	1.33	4.73	10.1	0.19	0.77
B. halodurans	528	VR	NR	149.52	NR	NR	0.28	NR	NR
C.saccharolyticus	1.03	14.3	33.5	99.9	68.1	118	100.32	4.756	3.525
D.turgidum	24.6	61.5	54.5	195	81	49	7.93	1.33	0.9
B. subtilis	53	NR	NR	153	NR	NR	2.88	NR	NR
T.composti	33.8	87.9	NR	19.77	2.14	NR	0.59	0.024	NR

[a]LR: L-Rhamnose; [b]DA: D-allose; [c]DR: D-ribose.

Crystal Structure of L-Rhamnose Isomerase

Korndorfer et al., successfully analyzed the crystal structure of L-RhI from *E. coli* (PDB accession number: 1d8w) by multiple isomorphic replacement methods for the first time in 2000. L-RhI from *E. coli* exists in the form of a compact tetramer in the crystal, and the apparent relative molecular weight is about 188 kDa. The single-subunit structure of *E. coli* L-RI consists of 420 amino acids, the subunit structure generally includes three functional domains: N-terminal domain, central domain, and C-terminal domain. The central domain is a (β/α)8-barrel core composed of 8 α-helices and 8 β-sheets areas. The N-terminal and C-terminal residues are randomly coiled into a small domain by several α-helices, which participate in the single interbody interactions and composition of active sites. Two sites that may be used for metal ion binding in the *E. coli* L-RhI structure. One is called "Zn^{2+}" which is considered to be the substrate-binding site of the enzyme to help stabilize the conformation of the substrate and the key amino acid residues involved in coordination are Glu234, Asp267, His294, and Asp334; Another "Mn^{2+}" point is considered as the catalytic site of the enzyme, which is used to help proton transfer to realize the catalytic process of the substrate and the key amino acid residues involved in coordination are His262 and Asp296. These key amino acid residues constituting the active center of L-RhI are highly conserved among the reported L-RhIs.

Meanwhile, the crystal structures of allose isomerases such as fucose isomerase (PDB accession number: 1FUI), triose phosphate isomerase (PDB accession number: 1TIM), glucose 6-phosphate isomerase (PDB accession number: 1PGI), phosphomannose isomerase (PDB accession number: 1PMI) and xylose isomerase (PDB accession number: 3XIM) have been successively reported. Through homology comparison, it was found that the crystal structure of *E. coli* L-RhI was the closest to D-xylose isomerase (D-XI), and the sequence similarity was 13%. It was speculated that the two were derived from the same precursor gene. The $(\beta/\alpha)_8$ barrel core regions of the two are very similar but different from D-XI, *E. coli* L-RhI has some strongly hydrophobic amino acid residues (Val53, Ile105, Tyr106, and Phe336) around the active center region, this special hydrophobic environment corresponds to the vicinity of C-6 of the substrate, which is presumed to be used for substrate recognition, which explains that the substrate affinity of *E. coli* L-RhI for L-rhamnose is higher than that of L- mannose.

Substrate Specificity of L-Rhamnose Isomerase

L-RhI has a broad substrate spectrum and can catalyze isomerization reactions including many pairs of five- and six-membered aldoses. The kinetics parameter of L-RhI from different sources on three substrates are compared in Table 4.

All reported microbial-derived L-RhIs have one thing in common, that is, they have a higher affinity for L-rhamnose and are more likely to catalyze the isomerization reaction of L-rhamnose to L-rhamnose. At the same time, there are certain differences in the specificity of L-RhI derived from different microorganisms to various substrates. For example, L-RhI derived from *E. coli* can catalyze L-rhamnose, L-lyxose, and L-mannose, but neither D-allose nor D-ribose can be used as substrates, while L-RhI derived from *P. stutzeri* has a broader substrate spectrum. Through the analysis of the crystal structure of L-RhI, Korndorfer and Yoshida et al. found that there is a special hydrophobic cavity formed by Val53, Ile105, Tyr106, and Phe336 in *E. coli* L-RhI to recognize the substrate, resulting in strict substrate specificity. While the substrate-binding site of *P. stutzeri* L-RhI is consistent with the binding form of substrate 1, 2, and 3 positions with *E. coli* L-RhI, and the binding form of substrate 4, 5, and 6 positions are closer to that of D-Xylose isomerase (D-XI) does not have a hydrophobic cavity, and the substrate only forms hydrophobic interactions with adjacent Phe66, so it has a wider substrate specificity.

Catalytic Mechanism of L-Rhamnose Isomerase

Based on the analysis of the crystal structure of *E. coli* L-RhI, Korndorferet al. speculated that the mechanism of negative hydrogen ion transfer mediated by metal ions may be the main reaction mechanism of L-RI catalyzing the isomerization of L-rhamnose to L-rhamnose. Yoshida et al. analyzed the crystal structure of L-RhI derived from *P. stutzeri* and studied the binding of the enzyme with the reaction products L-rhamnose and D-allose in the crystalline state to speculate its catalytic mechanism. The team further studied the catalytic process of *P. stutzeri* L-RhI through X-ray structural analysis and obtained the catalytic prediction process. They believe that there are two forms of catalytic sites of the L-RhI active center, namely AD type and BC type, and the exchange between the two forms is realized by metal ion displacement. Among them, the AD type is used to identify the five-membered ring ketose substrate. O-1, O-2, and O-3 of the substrate form coordination bonds with metal ions Mn1 and Mn2. Asp327 participates in the ring-opening of furan sugar and assists in the transfer of protons from O-2 to O-5. After ring opening, the catalytically active water molecule (W4) intervenes to assist the proton transfer from O-1 to O-2 and then attacks C-2 through the negative hydrogen ion (H^-) protected by trp179 to realize the isomerization reaction from a ketose to aldose. The BC type is used to identify the aldose substrate of the six-membered rings and the O-2 and O-3 of the substrate form a coordination bond with the metal ion Mn1. Different from the AD type, it is the catalytically active water molecule that participates in the ring-opening of the pyranose. After ring opening, the catalytic site is changed from BC type to AD type through the displacement of Mn2, and W4 is activated to intervene to assist the proton transfer from O-2 to O-1, and then H^- attacks C-1 to complete aldose to complete the isomerization reaction from aldose to ketose (Yoshida et al., 2006).

Biotransformation of D-Allose

Since D-allose is rare in nature, the research on the physiological function of D-allose and the promotion of the application of D-allose is limited. Therefore, large-scale production of D-allose becomes important. D-allulose biotransformation is the most suitable method for the production of D-allose, which can be obtained from D-allulose by using sugar isomerases such as L-rhamnose isomerase (L-RhI), D-ribose-5-phosphate isomerase (RPI) and D-galactose-6-phosphate isozyme isomerization (GPI). L-rhamnose isomerase is commonly used for D-allose biotransformation from D-allulose. However, using D-allulose as a substrate is not economically feasible for industrial applications. Han et al. reported that using a dual enzyme cascade of D-allulose 3-epimerase (DPE) and L-rhamnose isomerase (L-RhI) as bio-catalyst, 5.12 g/L D-allulose and 2.04 g/L of D-allose was obtained from 2% D-fructose as the substrate after 10 h reaction (Han et al., 2014a).

Ribitol Dehydrogenase (RDH) and Bioproduction of Allitol

Backgrounds

Allitol is a six-carbon rare sugar alcohol (Figure 4). It is a colorless crystalline solid, easily soluble in water, and has a sweet taste. It needs to be preserved at

2-8°C. Table 5 shows the properties of allitol. It is a non-hazardous substance according to the Globally Harmonized System of Classification and Labelling of Chemicals and Directive 67/548/EEC (Hassanin et al., 2016b).

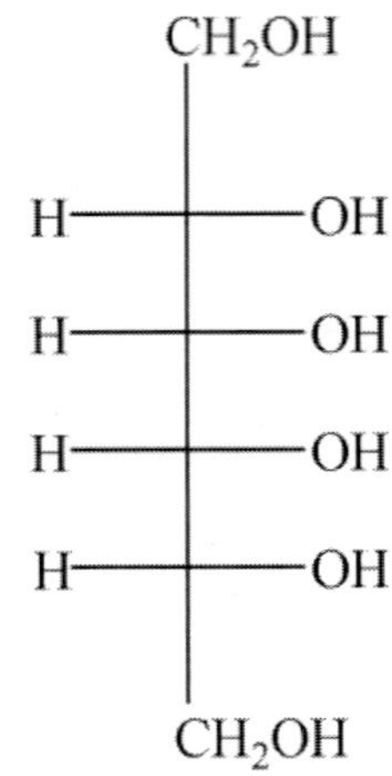

Figure 4. The structural formula for allitol.

Table 5. Properties of allitol

Property	Value
Molecular formula	$C_6H_{14}O_6$
Molecular weight	182.17
Density	1.596 g/cm^3
Melting point	152°C
Boiling point	494.9°C
Flash point	292.5°C

Allitol is a low-calorie rare sugar alcohol and can be used as a sweetener and bulking agent in the food industry (Hadipernata et al., 2012; Hassanin et al., 2016b). It also has some functional effects, like anti-obesity effects by suppressing lipid accumulation, and laxative effects used as a monosaccharide laxative for the treatment of constipation (Oosaka, 2009). Allitol can be used as an intermediate for the preparation of azasugars, which are medicines against diabetes, cancer, and viral infections, including AIDS (Hassanin et al., 2016b). Allitol can also be used as an anti-crystallization agent in the crystallization of sugar to prevent impurities from affecting the formation and growth of a sugar crystal nucleus (Hassanin et al., 2016b; Ottens et al., 2004). According to the strategy of Izumoring, allitol is located in the center of D/L-type rare sugar conversion and can be used for syntheses of other kinds of D/L-type rare sugars (such as D-allulose or L-allulose) (Izumori, 2006). Allitol can be biosynthesized from D-glucose, D-fructose, and D-allulose by multi-enzyme reaction and coenzyme regeneration system, respectively (Figure 5).

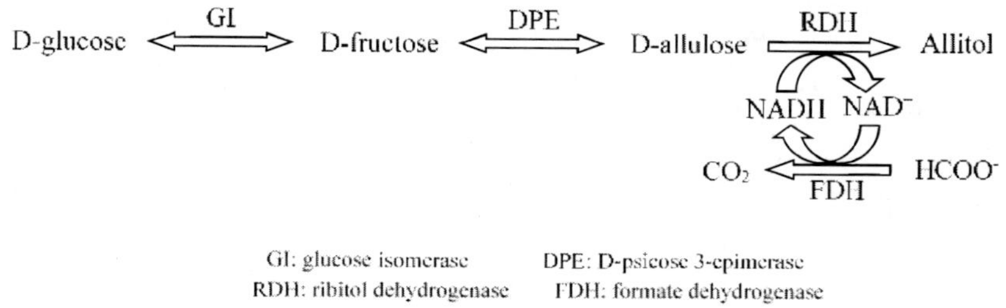

Figure 5. Scheme of allitol production from D-glucose, D-fructose and D-allulose.

Source and Properties of Ribitol Dehydrogenase

Allitol can be produced from D-allulose by Ribitol dehydrogenase (RDH) (Lu et al., 2019). Allitol was firstly produced from D-allulose by using the whole-cell biotransformation method using *Enterobacter agglomerans* strain 221E (Muniruzzaman et al., 1995b). In the same year, Muniruzzaman et al. purified the ribitol dehydrogenase derived from *Enterobacter agglomerans* strain 221E and studied its enzymatic properties (Muniruzzaman et al., 1995a) (Table 2). They found that it also can catalyze D-fructose to produce the byproduct of D-sorbitol. In 2000, Takeshita et al. used ribitol dehydrogenase from *Klebsiella pneumoniae* IF0 3321 to produce allitol by using a multi-enzyme reaction system and a coenzyme regeneration system to produce allitol from D-fructose and D-allulose, respectively (Takeshita et al., 2000). In 2015, Zhu et al. studied the physical and chemical properties of ribitol dehydrogenase from *Klebsiella oxytoca* and applied it to the production of allitol (Zhu et al., 2015). In 2016, Hassanin et al. reported that ribitol dehydrogenase from *Providencia alcalifaciens* RIMD 1656011 can catalyze D-allulose to produce allitol without producing any byproducts, and the enzymes have a high affinity for NAD^+ and are easy to carry out enzymatic reactions (Hassanin et al., 2016bc). Table 6 is a comparison of the enzymatic properties of ribitol dehydrogenase from different strains.

Table 6. Comparison of enzymatic properties of ribitol dehydrogenase from different strains

Conditions / Strains	pH	Temperature (℃)	K_m^{NAD+}	Substrate relative activities (%)		
				Ribitol	**Allitol**	**D-sorbitol**
E. agglomerans	9.0-10.0	50-60	2.36	100	66	5
K. pneumoniae	8.0	30	ND	100	11	0
K. oxytoca	8.0	40	ND	ND	ND	ND
P. alcalifaciens	10.0	35	0.042	100	8.6	0

ND, not detected.

Source and Properties of Formate Dehydrogenase for NADH Regeneration

Formate dehydrogenase can reduce NAD^+ to NADH, provide the coenzyme NADH for NADH-dependent oxidoreductase, and promote the smooth progress of the redox reaction (Tishkov et al., 2004; Wen et al., 2019). Formate dehydrogenase is widely present in bacteria (Nanba et al., 2003; Pometun et al., 2020; Rojkova et al., 1999), yeast (Labrou et al., 2001), fungi (Chow et al., 1993), higher eukaryotes and higher plants (Davison, 1951). According to reports, compared with yeast, formate dehydrogenase derived from bacteria has higher enzymatic activity and stable properties (Huang Zhihua et al., 2006). In 2015, Alekseeva et al. found that after mutating the alanine (Ala, A) at position 198 of the formate dehydrogenase from *Pseudomonas* sp. 101 to glycine (Gly, G), the thermal stability of the enzyme increased by 2.5, the K_m^{NAD+} value decreased from 60 to 35, the affinity of formate dehydrogenase and NAD^+ increased, and the enzyme activity increased (Table 7) (Alekseeva et al., 2015).

As a coenzyme regeneration system, formate dehydrogenase has the following advantages: first, the enzyme's substrate sodium formate or formic acid is cheap, and many enzymes have a high tolerance to sodium formate or formic acid; second, only one by-product CO_2 is produced during the reaction and it can easily overflow from the reaction system, which is conducive to the later separation and purification; third, the reaction process is irreversible, which is conducive to product production, etc. (Zhang et al., 2019).

Table 7. Kinetic parameters of native and mutant formate dehydrogenases from *Pseudomonas* sp. 101

Enzymes*	K_m^{NAD+}, μM	K_m^{HCOO-}, mM	K_{cat}, s^{-1}	K_{cat}/ K_m^{NAD+}, $mM^{-1}s^{-1}$
PseFDH	60 ± 5	6.5 ± 0.2	7.3 ± 0.2	122
PseFDH A198G	35 ± 2	7.5 ± 0.2	7.3 ± 0.1	209

* Alekseeva et al., 2015. Pse, *Pseudomonas* sp. 101; FDH, formate dehydrogenase; A, Ala; G, Gly.

Allitol Biotransformation Using Enzymatic Method

Allitol Bioproduction from D-Allulose

In 2000, Takeshita et al. successfully realized the production of allitol from D-allulose by using RDH and FDH crude enzyme solution from *Klebsiella pneumoniae* IF0 3321 (Takeshita et al., 2000). In 2016, Hassanin et al. reported that RDH from *Providencia alcalifaciens* RIMD 1656011 not only has the function of converting ribitol to D-ribulose but also catalyzes D-allulose to allitol (Hassanin et al., 2016c). In the same year, the RDH crude enzyme solution and FDH crude enzyme solution from yeast *Ogataea parapolymorpha* DL-1were used to produce allitol from D-allulose and obtained allitol crystal with 95% purity (Hassanin et al., 2016a).

Allitol Bioproduction from D-Fructose

While Takeshita et al., successfully realized the production of allitol from D-allulose by multi-enzyme reaction, they also added DTE enzyme from *Pseudomonas cichorii* ST-24 which can catalyze D-fructose to D-allulose, and then D-allulose was catalyzed to allitol by RDH and FDH enzymes (Takeshita et al., 2000).

Ribitol dehydrogenase catalyzes the biotransformation of D-allulose into allitol, which can break the thermodynamic equilibrium of the catalyzed reaction by DPE or DTE, and make the substrate D-fructose move towards the formation of D-allulose, thus increasing the yield of allitol. Although the enzyme catalytic method can directly produce allitol from D-allulose and D-fructose, the concentration of substrate used is low, the reaction time is long, the yield of allitol is not high, and the preparation of enzyme is time-consuming and labor-consuming. In addition, the addition of initial coenzyme NAD^+/NADH increases the production cost, which is not conducive to the industrial production of allitol.

Allitol Biotransformation Using Microbial Method

Allitol Bioproduction from D-Allulose

As early as 1995, Muniruzzaman et al. realized the production of allitol from D-allulose using *Enterobacter agglomerans* strain 221e (Muniruzzaman et al., 1995b). In 2014, Han et al. used *Klebsiella oxytoca* G4A4 to produce allitol from D-allulose (Han et al., 2014b). In 2018, Hassanin et al. constructed a recombinant strain capable of expressing both RDH (*Providencia alcalifaciens* RIMD 1656011) and FDH (*Ogataea parapolymorpha* DL-1) enzymes and used this strain to catalyze D-allulose to allitol (Hassanin et al., 2018). In 2022, a recombinant strain capable of expressing both RDH (*Providencia alcalifaciens* RIMD 1656011) and FDH (*Pseudomonas* sp. 101) was constructed and used for whole-cell biotransformation. The conversion rate was 64.3% when the substrate concentration was 90 g/L (Wen et al., 2022). In order to overcome the diffusion limitation of the intermediates of biotransformation reactions and avoid the interference of the intermediates on the normal metabolism of cells, Lin et al. had also invented a triangular mesh RNA scaffold for intracellular recombinant protein immobilization, which combines RDH and FDH. Anchoring in adjacent spaces makes the recombinase molecules in the artificial synthesis pathway close to each other in space, forming an artificial synthesis pathway, promoting the efficient synthesis of allitol, and reducing the interference to the normal metabolism of the host bacteria. When using the recombinant strain with RNA scaffold anchors RDH and FDH as a catalyst to transforme D-allulose into allitol, the yield of allitol was significantly improved (Lin, et al., 2020ab).

Allitol Bioproduction from D-Fructose

In 2015, Zhu et al. constructed a recombinant *E. coli* BL21 star (DE3)-pCDF-*rdh-fdh*/pET-*dpe*/pACYC184 M-*glf* which simultaneously expressed DPE, RDH, FDH, and GLF (glucose/fructose transport membrane protein) enzymes, yielded 48.62 g/L allitol from 500 mM D-fructose (Zhu et al., 2015). In 2020, Wen et al. constructed a two-plasmid recombinant strain capable of simultaneously expressing DPE, RDH, and FDH, and obtained a large number of bacterial cells by fed-batch culture, which was used to convert D-fructose into allitol. When the concentration of D-fructose was 100 g/L, the conversion rate of allitol reached 63% (Wen et al., 2020b). Considering the poor stability of dual-plasmid strains, and two antibiotics were needed to cultivate dual-plasmid strains, which increased the production cost. Wen et al. constructed a single-plasmid recombinant strain that can express DPE, RDH, and FDH at the same time. Whole-cell biotransformation was carried out by using this strain and the same allitol conversion rate of 63% was obtained. Finally, allitol crystals of high purity were obtained (Wen et al., 2020a). Compared to the dual-plasmid strains, the stability of the single-plasmid recombinant strain is better, and only one antibiotic needs to be added during the strain culture process, which greatly improves the process stability and reduces the production cost.

Allitol Bioproduction from D-Glucose

Although the cost of D-fructose is lower than that of D-allulose, it is still significantly higher than that of D-glucose. Glucose isomerase can efficiently catalyze the reversible transformation between D-glucose and D-fructose. Therefore, the introduction of glucose isomerase into the allitol production process can realize the bioconversion of allitol from D-glucose. In 2020, Wen et al., use for the first time an immobilized glucose isomerase and recombinant strain capable of simultaneously expressing DPE, RDH, and FDH as mixed catalysts to produce 12.7 g/L allitol with 50 g/L D-glucose as substrate, which established a feasible technical route for large-scale and low-cost production of allitol (Wen et al., 2020b).

In summary, the biotransformation of D-glucose, D-fructose, and D-allulose to produce allitol can be achieved by an enzymatic method and microbial biotransformation method, respectively. However, compared with the microbial biotransformation method, the enzymatic transformation process is more complicated. First, the extraction, separation, purification, and immobilization of the enzyme were needed, and easily leads to the loss of the enzyme; secondly, the enzymatic method requires the addition of the expensive coenzyme NAD^+/NADH. The above reasons lead to a complex production process and increased production cost, which is not conducive to the industrial production of allitol. The production of allitol by the microbial transformation method can effectively avoid the defects of the enzymatic method. The microbial cells with high cell catalytic activity can be easily obtained by centrifugation, and the intracellular coenzyme NADH can be used as a substrate. It does not need to add additional coenzyme, which can greatly save the production cost. In addition, it is also convenient for separation and purification in the later stage, which is beneficial to the industrial production of allitol.

Enzymes and Bioproduction of L-Xylulose, L-Xylose and L-Lyxose

Backgrounds

In recent years, interest in L-rare sugars and nucleosides derived from them has attracted attention in medicine. L-Rare sugars hold enormous potential for practical applications. They are often recognized as important precursors of pharmaceutical drug molecules (Frihed et al., 2015). L-rare sugars have been widely used as antiviral drugs in the treatment of severe viral diseases, such as those caused by the HIV or hepatitis virus, while they also show promise as anticancer and cardioprotective agents (Li et al., 2013). Glycosidase and glucose reabsorption inhibition effects have also been reported for L-rare sugars and their derivatives. L-Xylulose, L-xylose, and L-lyxose are among these valuable L-forms of rare sugars.

L-Xylulose has been shown a specific inhibitory of glycosidases, thus, it can be used for the treatment of diabetes. An L-xylulose-containing containing drug preparation for reducing blood sugar levels in humans and animals has been patented (Heinz et al., 1998), as well it is a reliable indicator of patients with acute or chronic hepatitis or liver cirrhosis (Hiroshi et al., 1976; Muniruzzaman et al., 1996). It has also been reported that the occurrence of L-xylulose in the urine can be used to indicate the pentosuria caused by L-xylulose reductase deficiency in the oxidative pathway of glucuronate metabolism (Pierce et al., 2011). L-xylulose can also be used as a starting material to produce other important rare sugars such as L-xylose, L-lyxose, and L-ribulose (Fang, 2018). L-Xylose is an aldopentose that has a molecular formula $C_5H_{10}O_5$ and a molar mass of 150.13 g/mol. L-Xylose is rare in nature. L-Xylose is a suitable starting material for the production of L-ribonucleosides, polyhydroxypyrrolidines, and related analogs. These compounds are shown to have anti-HIV effects, inhibit tumor growth, and act as α- and β-glucosidase inhibitors, which are of relevance for the development of diabetes drugs (Behr et al., 2007). L-Lyxose is related to L-rhamnose or to the hexose L-mannose structurally. L-Lyxose is used as an essential precursor in the preparation of avilamycin A (Hofmann et al., 2005). It is also used as an L-fucosidase inhibitor that could be used as a potential anti-inflammatory treatment (Chevrier et al., 2006). However, the limited availability of these valuable compounds restricts their wide applications as well as the discovery of other novel biological activities. Both chemical and biological approaches are used for L-xylulose, L-xylose, and L-lyxose synthesis. While the tedious chemical synthesis does not satisfy the increasing demand for these rare sugars, biological synthesis (both enzymatic and microbial biotransformation) has emerged as a highly promising alternative.

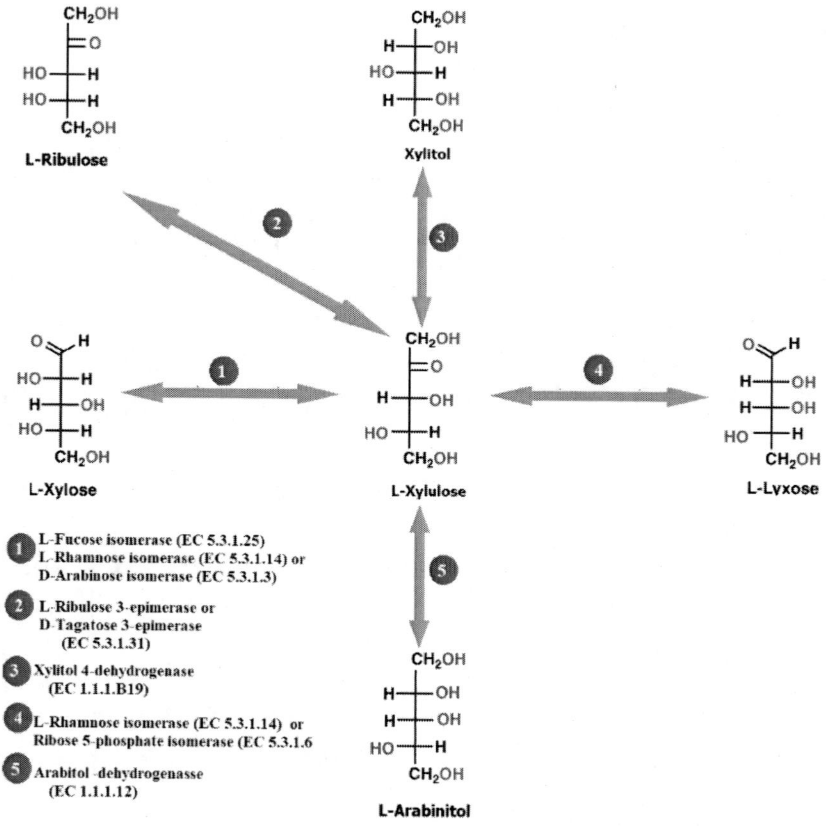

Figure 6. L-Xylulose bioconversion based on Izumoring strategy.

Enzymes and Bioproduction of L-Xylulose

Recently, biological methods are becoming widely applied in rare sugar production because of their simplicity, controllability, high stereo-selectivity, and ecological safety compared to chemical methods. A systematic enzymatic strategy for rare sugar bioproduction has been established, named the Izumoring strategy that connects all monosaccharides cyclically transformed by ketose C-3 epimerization, aldose-ketose isomerization, and monosaccharide oxidization-reduction (Izumori, 2006). According to the Izumoring strategy, L-xylulose can be produced from the isomerization of L-lyxose and L-xylose by pentose isomerases, the epimerization of L-ribulose by ketose 3-epimerase, and the oxidation of xylitol and L-arabinitol by oxidoreductases (Figure 6).

L-Xylulose Bioproduction from L-Lyxose

L-Rhamnose isomerase (L-RhI, EC 5.3.1.14) is an aldose isomerase reversibly catalyzing the isomerization between L-rhamnose and L-rhamnulose. In addition, L-RhI exhibits broad substrate specificity toward various aldoses and ketoses and thus displays enormous potential in the biological production of many expensive rare sugars (Leang et al., 2004a). L-RhIs have been characterized from various microorganisms, including *Pseudomonas stutzeri* (Leang et al., 2004b), *Bacillus pallidus* Y25 (Poonperm et al., 2007b), *Thermoanaerobacterium saccharolyticum* NTOU1 (Lin et al., 2010), *Thermotoga maritima* ATCC 43589 (Park et al., 2010), *Caldicellulosiruptor saccharolyticus* ATCC 43494 (Lin et al., 2011), *B. halodurans* ATCC BAA-125 (Prabhu et al., 2011), *Mesorhizobium loti* Tono (Takata et al., 2011), *Dictyoglomus turgidum* DSMZ 6724 (Kim et al., 2013), *B. subtilis* ATCC 23857 (Park, 2014), *B. subtilis* str. 168 (Bai et al., 2015) and *Clostridium stercorarium* (Min et al., 2017). All the reported L-RhIs showed the optimum substrate to be L-rhamnose. L-Lyxose was the second favored substrate, and the enzyme from *T. saccharolyticum* NTOU1 showed much higher specific activity (130 U mg^{-1}) toward the isomerization of L-lyxose to L-xylulose than other L-RhIs (Lin et al., 2010). The L-RhI from *B. subtilis* ATCC 23857 (Park, 2014) and *T. maritima* ATCC 43589 (Park et al., 2010) catalyzed the isomerization of L-lyxose and L-xylulose and the final equilibrium ratio between L-lyxose and L-xylulose was 40:60 and 45:55, respectively. However, the expensive price of L-lyxose makes it inappropriate for L-xylulose preparation.

L-Xylulose Bioproduction from L-Xylose

L-Xylose is an expensive rare sugar that shows great application potential in the pharmaceutical and healthcare fields. L-xylulose could be produced by the isomerization of L-xylose using isomerases such as D-arabinose isomerase (D-AI, E.C 5.3.1.3), L-fucose isomerase (L-FucI, E.C 5.3.1.25), and L-RI. The D-AI from *K. pneumoniae* shows a specific activity toward L-xylose of 4.7 U mg^{-1} (Menavuvu et al., 2006). Cell-free *P. stutzeri* L-RI catalyzes the isomerization between L-xylose, L-xylulose, and L-lyxose with an equilibrium ratio of 61:35:4 (Leang et al., 2004a), and the immobilized *P. stutzeri* L-RI on chitopearls shows an equilibrium ratio of 26:53:21 (Granstrom et al., 2005). L-FucI from *E. coli* favors the formation of L-xylose and catalyzes the isomerization of L-xylulose and L-xylose with a reaction equilibrium of 15:85 (Usvalampi et al., 2012). However, L-xylose is expensive, and the production of L-xylulose from L-xylose makes it inappropriate for commercial production.

L-Xylulose Bioproduction from L-Ribulose

L-Rbulose is the epimer of L-xylulose at the C-3 position. L-Xylulose could be obtained from L-ribulose through C-3 epimerization catalyzed by ketose 3-epimerase (Meng et al., 2016). In 1994, the first ketose 3-epimerase was reported as D-tagatose 3-epimerase (D-TE, EC 5.1.3.31) from *Pseudomonas cichorii* (Itoh et al., 1994). It exhibits C-3 epimerization activity toward a variety of ketopentose and ketohexoses with the highest activity toward D-tagatose. Compared to D-tagatose activity, *P. cichorii* D-TE shows 20% and 70% of relative activity toward L-xylulose and L-ribulose, respectively. The equilibrium ratio between L-ribulose and L-xylulose is 30 and 70 respectively (Itoh et al., 1994). In 2013, Uechi et al. reported L-ribulose-3-epimerase (L-RE, EC 5.1.3.31) from *Mesorhizobium loti* (Uechi et al., 2013) with an optimum ketopentose substrate specificity. The specific activities toward L-ribulose and L-xylulose are 230 and 96 U mg^{-1} respectively. The equilibrium ratio of L-ribulose to L-xylulose is calculated to be 40:60. More recently, a novel thermostable L-RE from *Labedella endophytica* was successfully overexpressed and characterized by Chen et al., (2021), and its optimum substrate is L-ribulose with a specific activity of 172.8 U mg^{-1}. So far, five L-ribulose 3-epimerases have been identified and characterized including those from *Mesorhizobium loti* MAFF303099 (Uechi et al., 2013), *Thermotoga maritima* MSB8 (Shin et al., 2017), *Arthrobacter globiformis* M30 (Yoshihara et al., 2017), *Methylomonus* sp. DH-1 (Yoshida et al., 2021), and *Labedella endophytica* (Chen et al., 2021). All the reported L-REs showed their optimum substrate as L-ribulose, except for the L-RE of *T. maritima* in which D-erythrose is its optimum substrate followed by L-ribulose (Table 8). At present, the prices of both L-ribulose and L-xylulose are extremely high. Therefore, it is not economically feasible to produce L-xylulose from L-ribulose catalyzed by D-TE or L-RE.

Table 8. Biochemical properties of L-RI from various microbial sources

L-RI Source	Substrate	Product	Specific activity	Optimum			Reference
				pH	Temperature (°C)	Metal ion	
				8.0	60	Mn^{2+}	
M. loti	L-Xylulose	L-Ribulose	96				Uechi et al., 2013
	D-Allulose	D-Fructose	55				
	D-Tagatose	D-Sorbose	5.3				
	D- Fructose	D-Allulose	4				
T. maritima	D-Erythrose	D-Threose	18.7	7.0	80	Mn2+	Shin et al., 2017
	L-Ribulose	L-Xylulose	5				
	D-Threose	D-Erythrose	4.1				
	L-xylulose	L-Ribulose	2				
	D-Tagatose	D-Sorbose	1				
	D-Allulose	D-Fructose	0.1				
	D-Fructose	D-Allulose	0.1				
A. globiformis	L-Ribulose	L-Xylulose	121.3	8.0	70	Mg2+	Yoshihara et al., 2018
	D-Allulose	D-Fructose	77				
	D-Ribulose	D-Xylulose	21.4				
	L-Xylulose	L-Ribulose	18				
	D-Xylulose	D-Ribulose	10.2				
M. sp.	L-Ribulose	L-Xylulose	82.1	NR	70	NR	Yoshida et al., 2021
	D-Allulose	D-Fructose	53.3				
	D-Fructose	D-Allulose	47.8				
	L-xylulose	L-Ribulose	25.6				
	D-Tagatose	D-Sorbose	13.6				
	L-Sorbose	L-Tagatose	7.5				
L. endophytica	L-Ribulose	L-Xylulose	172.8	6.0	80	Ni2+	Chen et al., 2021
	D-Fructose	D-Allulose	110.7				
	L-xylulose	L-Ribulose	108.7				
	D-Allulose	D-Fructose	107.3				
	D-Tagatose	D-Sorbose	46.2				
	L- Sorbose	L-Tagatose	19.6				

L-Xylulose Bioproduction from L-Arabinitol

L-Xylulose can be alternatively obtained from L-Arabinitol oxidized by Arabitol-dehydrogenase (ArDH, EC 1.1.1.12). Richard et al. first cloned the ArDH from *Trichoderma reesei* in *S. cerevisiae* and confirmed the enzyme's ability to produce L-xylulose from L-arabitol using a cell extract (Richard et al., 2001). Later, other ArDHs from *Aspergillus oryzae* (Mahmud et al., 2013) and *T. reesei* (Tiwari et al., 2014) were identified and showed L-xylulose-producing ability in the presence of NADH. Recombinant ArDH from *T. reesei* displayed a high catalytic efficiency in which, after a 2-h reaction at pH 9.5, the equilibrium conversion ratio from L-arabitol (5 g/L) to L-xylulose was

86 ± 3% and the volumetric productivity was 2.2 g/(L·h). Furthermore, Singh et al. (2014) applied ArDH immobilized on aldehyde functionalized silicon oxide nanoparticles to produce L-xylulose fromL-arabitol, and finally achieved 66% conversion and 7.9 g/(L·h) productivity. Recently, repetitive production of L-xylulose from L-arabitol by an immobilized whole-cell biocatalyst harboring ArDH coupled with a cofactor recycling system was proposed by Gao et al. In the system, the efficiency of L-xylulose production was increased to 96%, and the immobilized whole cells could retain 65% of the yield after 7 cycles (Gao et al., 2015) using 150 mM of L-arabinitol. Gao et al. (2016) also reported that using a cell-free biocatalytic system containing ArDH for arabinitol oxidation and NADH oxidase for cofactor regeneration, a conversion of 78.4% was achieved under optimized reaction conditions containing 250 mM L-arabinitol and 8.2 mM NAD^+ within 8 h. However, the high cost of L-arabitol is a limiting factor of this approach for industrial application.

L-Xylulose Bioproduction from L-Xylitol

The more readily available, less expensive, and structurally similar sugar alcohol xylitol can serve as an ideal substrate for L-xylulose production. As an intermediate of xylitol metabolism, L-xylulose formation has been reported in *Pantoea ananatis* (Doten et al., 1985), *Alcaligenes* sp. (Khan et al., 1991), and *Bacillus pallidus* (Poonperm et al., 2007a). Further investigation demonstrated that an NAD^+-dependent xylitol-4-dehydrogenase (XDH, EC 1.1.1.B19) mediated this interconversion (Aarnikunnas et al., 2006), and its primary structure is similar to members of the short-chain dehydrogenase/reductase family (Persson et al., 2013). XDH naturally exists in humans (El et al., 2010), rodents (Ishikura et al., 2001), fungi (Yamasaki et al., 2017), and bacteria (Doten et al., 1985a). So far, the following strains of bacteria are reported to produce L-xylulose via XDH-catalyzed conversion of xylitol (Table 9), including *Alcaligenes* sp. (Khan et al., 1991), *Pantoea ananatis* (formerly named *Erwinia uredovora*), (Doten et al., 1985c; Aarnikunnas et al., 2006; Usvalampi et al., 2009), *Bacillus pallidus* (Poonperm et al., 2007a; Takata et al., 2010), *B. megaterium* (Zhang et al., 2014) and *Erwinia aphidicola* SK47.001 (Li et al., 2021). All of them were reported as NAD^+ dependent enzymes.

Table 9. Summary of L-xylulose production from xylitol or L-arabinitol using XDH or ArDH enzymes

Substrate (g/L)	Biocatalyst	Reaction time (h)	L-Xylulose (g/L)	Conversion rate (%)	References
Xylitol (5)	Resting cells of Alcaligenes sp. strain 701B	24	4	80	Khan et al., 1991
Xylitol (5)	Resting cells of P. ananatis ATCC 43074	18	3.5	70	Doten and Mortlock 1985
Xylitol (10)	Resting E. coli cells expressing XDH of P. ananatis	19	7	70	Aarnikunnas et al., 2006
Xylitol (20)	Resting cells of Bacillus pallidus Y25	12	16	85	Poonperm et al., 2007a
Xylitol (20)	B. megaterium ZN-14	24	5.3	26.62	Zhang et al., 2014
Xylitol (20)	Resting E. coli cells expressing XDH of B. pallidus Y25	48	10.6	53	Takata et al., 2010
Ribitol (20)	Resting cells of Acetobacter aceti and immobilized purified recombinant L-Ribulose-3-epimerase of Mesorhizobium loti onto Chitopearl Beads BCW2510	16	7.2	36	Uechi et al., 2013
Xylitol (30)	Immobilized recombinant XDH of Rhizobium etli CFN42 on activated silicon nanoparticles	2		71	Zhang et al., 2011
L-Arabinitol (30)	Purified recombinant L-Arabinitol dehydrogenase form H. jecorina	2	3.9	86	Tiwari et al., 2014
L-Arabinitol (30)	Immobilized recombinant HjLAD on aldehyde-functionalized silicon oxide nanoparticles	2.5		66	Singh et al., 2014
L-Arabinitol (22.8)	Whole E. coli cells separately expressing HjLAD and NADH oxides of S. pyogenes mixed in equal ratio and immobilized on sodium alginate	8	21.9	96	Gao et al., 2016
Xylitol (22)	Growing E. coli cells harboring XDH of P. ananatis and NADH oxidase S. pneumoniae for cofactor regeneration	80	14	65	Han et al., 2017

Substrate (g/L)	**Biocatalyst**	**Reaction time (h)**	**L-Xylulose (g/L)**	**Conversion rate (%)**	**References**
Xylitol (20)	Resting E. coli cells harboring XDH of P. ananatis grown on optimized media containing molasses and corn steep liquor	19	19.3	96.6	Tesfay et al., 2021b
Xylitol (50)	Resting E. coli cells harboring XDH of P. ananatis and NADH oxidase from S. mutans for cofactor regeneration grown on optimized media containing molasses and corn steep liquor	20	48.45	96	Tesfay et al., 2021a
Xylitol (3)	Purified recombinant ArDH from Aspergillus nidulans coupled with NADH oxidase of S. mutans	8	Not reported	84.6	Zhu et al., 2021

At a xylitol concentration of 5 g/L the highest L-xylulose yields from xylitol of 80% for *Alcaligenes* sp. strain 701B (Khan et al., 1991) and 70% for *P. ananatis* ATCC 43074 (Doten et al., 1985c) were obtained at 24 h and 18 h, respectively. Poonperm et al., (2007a) reported that maximum product yields under optimized conditions of 85% were achieved at xylitol concentrations of 20 g/L after 12 h of reaction using *B. pallidus*. The authors state that the conversion was carried out at low cell density since the L-xylulose was consumed by the cells at higher densities and trace amounts of L-xylose and L-arabitol could be observed. Cloning of XDH genes from *B. pallidus* and *P. ananatis* was also reported, and the recombinant strains were used to produce L-xylulose. With the recombinant cells of *E. coli* expressing XDH of *P. ananatis*, a 70% yield was achieved in 19 h, with an initial xylitol concentration of 10 g/L, whereas with *Alcaligenes* sp. strain 701B and *P. ananatis*, only 50% and 48% yields, respectively, could be achieved at this xylitol concentration. Using recombinant XDH from *B. pallidus*, only 51% conversion was achieved from 10 g/L of xylitol. More recently, through optimization of culture media (containing molasses and corn steep liquor) and the production conditions (pH, temperature, initial xylitol, and cell concentration), 96% conversion was achieved from 20 g/L xylitol using high *E. coli* cell concentration expressing XDH of *P. ananatis* (Tesfay et al., 2021a).

The possibility of exploiting the high yield of XDH in microorganisms has great application potential. L-Xylulose producing XDHs from xylitol were reported as NAD^+ dependent enzymes.

The continuous availability of NAD^+ provides a driving force for the biotransformation of NAD^+-dependent enzymes. The specific conversion of xylitol to L-xylulose involves the formation of NADH; hence, the persistence of NAD^+ is necessary to drive this process. Han et al., (2017) studied growing cells of recombinant *E. coli* co-expressing XDH with water-forming NADH oxidase (NOX) from *Streptococcus pneumoniae* in two plasmid systems at a 1 L scale using 5 g/L substrates and achieved 69% conversion of xylitol to L-xylulose. A water-forming NADH oxidase from *Streptococcus mutans* was co-expressed in *E. coli* with XDH of *Pantoea ananatis*, under optimized reaction conditions and high cell concentration, the co-expression system increased the intracellular NAD^+ concentration and achieved 48 g/L of L-xylulose was produced from 50 g/L of Xylitol (Tesfay et al., 2021b).

A recently identified XDH from *E. aphidicola* was co-expressed with water-forming NADH oxidase in food-grade *B. subtilis* strains in which the XDH activity was increased by 3.4-fold higher than the free enzyme (Li et al., 2021). Moreover, a new ArDH from *Aspergillus nidulans* cloned and expressed in *E. coli* and used to produce L-xylulose from 20 mM of xylitol coupled with NADH oxidase for NAD^+ regeneration, and achieved a conversion of 84.6% (Zhu et al., 2021). This could be an alternative approach to producing L-xylulose using ArDH enzymes in industries because xylitol is cheaper than L-arabinitol. At present, compared with other approaches, xylitol bioconversion is the most economical and favorable method for L-xylulose production.

L-Lyxose Bioproduction

L-Lyxose has been prepared via biotransformation and enzymatic reactions (Table 10). Bhuiyan et al. (1998) reported the enzymatic biotransformation of L-lyxose from ribitol, which involves oxidation of ribitol to L-ribulose by *Acetobacter aceti,* epimerization of L-ribulose to L-xylulose by D-tagatose 3-epimerase, and isomerization of L-xylulose to L-lyxose by rhamnose isomerase. Using a flask reaction, 5 g of L-lyxose crystal was recovered from 10 g of ribitol. A recombinant *P. stutzeri* L-RI was used to produce L-lyxose from L-xylulose, which came from the oxidation of xylitol by *A. species* 701B. The immobilized recombinant *P. stutzeri* L-RI produced 4.06 g L^{-1} L-lyxose and 4.94 g L^{-1} L-xylose from 19.2 g L^{-1} L-xylulose (Granstrom et al., 2005). Moreover, L-lyxose production from L-xylulose was reported using L-RI from *B. subtilis* ATCC 23857 (Park, 2014) and *T. maritima* ATCC 43589 (Park et al., 2010). The final equilibrium ratio between L-lyxose and L-xylulose was 40:60 for the L-RI of *B. subtilis* ATCC 23857 and 45:55 for the L-RI of *T. maritima* ATCC 43589 without any side product. Ribose-5-phosphate isomerase (Rpi, EC 5.3.1.6) has recently been used to produce several L-form rare sugars, including L-rhamnulose, L-lyxose, and L-tagatose (Chen et al., 2020). Rpi from *S. pneumoniae* ATCC BAA-255 was able to produce 350 g L^{-1} L-lyxose from 500 g L^{-1} L-xylulose with a volumetric productivity of 117 g L^{-1} h^{-1} and a conversion yield of 70% (Park et al., 2011).

Table 10. L-Lyxose production using different biocatalysts

Enzyme	Sources	Reaction condition (pH, Temperature)	Substrate (g/L)	L-Lyxose produced (g/L)	Equilibrium ration	References
L-RI	*B. subtilis*	8.0, 60°C	L-Xylulose (100)	40	60:40[a]	Park 2014
	T. maritima	8.0, 85°C	L-Xylulose (500)	225	55:45[a]	Park et al., 2010
	P. stutzeri	Not reported	L-Xylose	NR	61:35:4[b]	Leang et al., 2004
Immobilized L-RI	*P. stutzeri*	9.0, 42°C	L-Xylulose (19.2)	4.09	26:53:21[b]	Granstrom et al., 2005
Rpi	*S. pneumoniae*	7.5, 35°C	L-Xylulose (500)	350	30:70[a]	Park 2011

[a] = L-Xylulose: L-Lyxose; [b] = L-Xylulose: L-Xylose:L-Lyxose; L-RI =L-rhamnose isomerase; Rpi = Ribose-5-phosphate isomerase.

L-Xylose Bioproduction

Both chemical and biological methods of L-xylose production have been reported. L-Xylose has been prepared from D-glucose (Dimant et al., 1960) or D-gluconolactone (Yang et al., 2002) via chemical routes which were tedious and suffered from low yields. L-Xylose can be also produced from L-xylulose by isomerization. This can be achieved either enzymatically or chemically under alkaline conditions. In the chemical isomerization of L-xylulose to L-xylose, an aqueous solution of L-xylulose is incubated at 40°C for about 1 and a half days in the presence of NaOH and 18% ofL-xylose was produced (Heikkilä et al., 2002). Jokela et al. (2002) reported enzymatic production of L-xylose using xylose isomerase from *Streptomyces rubiginosus.* However, L-lyxose would be formed as a side product and the reaction was slow. As mentioned above in L-lyxose production, Granström et al. (2005) used immobilized L-rhamnose isomerase from *Pseudomonas stutzeri to* produce L-xylose from L-xylulose. L-Lyxose was formed as a side product and the final concentrations were 53:26:21 for L-xylulose, L-xylose, and L-lyxose, respectively. L-Fucose isomerase from *E. coli* was also used to catalyze the conversion and no L-lyxose was detected as a byproduct in this case. The optimal pH for this enzyme was alkaline and the ratio of L-xylulose/L-xylose was 15:85 at reaction equilibrium, which suggested that the formation of L-xylose was favored (Usvalampi, et al., 2012).

Isomerase and Dehydrogenase for Bioproduction of L-Ribose

Backgrounds

Ribose is a five-carbon sugar with two configurations, D-ribose and L-ribose (Figure 7), where L-ribose is extremely difficult to find in nature. L-Ribose is a white or off-white powder crystal, easily soluble in water and its solubility can reach 100 mg/ml. It is also soluble in ethanol and methanol, but insoluble in ether (Huang et al., 2019). Other physicochemical properties of L-ribose are shown in Table 11.

L-Ribose can be used as a precursor for the production of L-nucleoside analogs used for the syntheses of antiviral and antitumor drugs (Wang et al., 1998). Table 12 is some nucleoside drugs based on L-ribose (Huang et al., 2019). According to data released by UNAIDS, 37.7 million people worldwide are still infected with a virus, and the patients require lifelong medication. Therefore, the demand for antiviral drugs synthesized from L-ribose and its derivatives has increased accordingly.

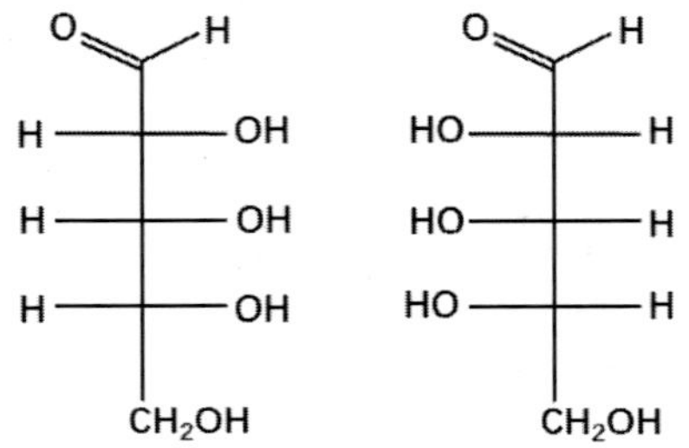

Figure 7. The structural formula for D-Ribose and L-Ribose.

Table 11. Properties of L-Ribose

Name	**L-Ribose**
Molecular formula	$C_5H_{10}O_5$
Synonyms	L-(+)-Ribose, (2S,3S,4S)-2,3,4,5-tetrahydroxypentanal
Molecular weight	150.13
Melting point	81-82 °C
Boiling point	191.65°C
Density	1.1897
Specific rotation	20 °C
Solubility	Soluble in water (100 mg/ml)

Table 12. Nucleoside analog drugs with L-ribose as the main structure

Drug name	**Application**
Telbivudine	Hepatitis B treatment, HBV
Emtricitabine	AIDS treatment, HIV; Hepatitis B treatment, HBV
Clevudine	Hepatitis B treatment, HBV
Levovirin	Hepatitis C treatment, HCV
Amdoxovir	AIDS treatment, HIV
Maribavir	anti-cytomegalovirus, HCMV
Troxacitabine	Leukemia treatment

Isomerase for Biotransformation of L-Ribose

L-Ribose can be obtained from L-ribulose by one-step isomerization, or from L-arabinose by two-step isomerization (Figure 8). L-Ribose production by using L-arabinose isomerase (L-AI) and L-ribose isomerase (L-RI) has a conversion yield of 20% (Helanto et al., 2009), while the conversion yield by using L-AI and mannose-6-phosphate isomerase (MPI) is about 33% (Kim et al., 2014).

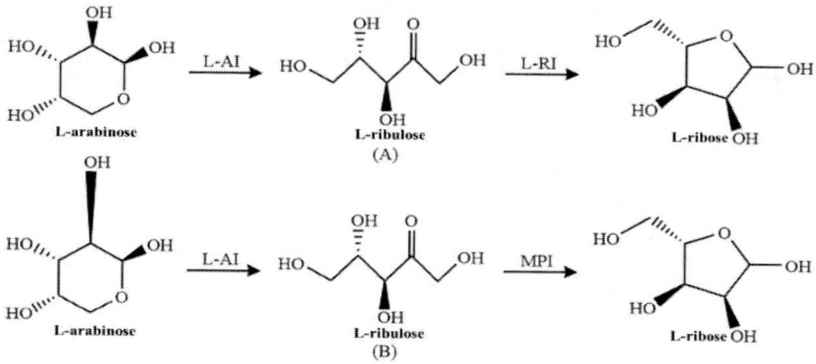

Figure 8. L-ribose isomerization synthesis strategy (L-AI, L-arabinose isomerase; L-RI, L-ribose isomerase; MPI, mannose-6-phosphate isomerase).

Enzyme Properties of L-RI and MPI

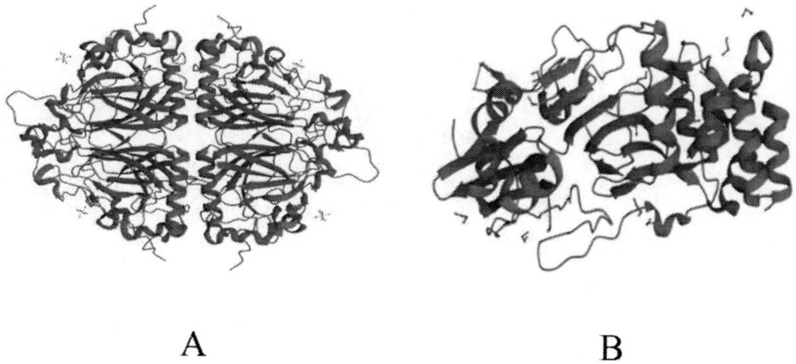

Figure 9. Structures of L-ribose isomerase(A) and mannose-6-phosphate isomerase(B).

Table 13. Comparison of the enzymatic properties of L-RI and MPI

Properties	**L-AI (*Acinetobacter sp. DL-28*)**	**MPI (*G. thermodenitrificans*)**
Optimal temp. (°C)	30-40	70
Optimal pH	9.0	7.0
Stable pH range	7.0-9.0	NR
M.W. (kDa)	120 (tetramer)	36
Substrate	L-ribose	L-ribose
Conversion. (%)	20	33
Relative thermal stability	unstable	stable

The optimal temperature of MPI was 70°C (Shimonishi et al., 1996), and the half-life is 27 h at 70°C. While L-RI (*Af*RI) from *Acidaminococcus fermentans* has an activity half-life of 7.2 h at 45°C (Xu et al., 2016ac). Some of the other properties are shown in Table 13 (Xu et al., 2016b). The thermal stability of MPI is higher than that of L-RI, the reason may be that L-RI is a tetrameric structure. The structures of L-ribose isomerase and mannose-6-phosphate isomerase were shown in Figure 9A and B respectively (Yoshida et al., 2014; Sagurthi et al., 2009).

Catalytic Mechanism of L-RI and MPI

The catalytic mechanism of L-RI and MPI in the biosynthesis of L-ribose is a cis-enediol intermediate mechanism. In this mechanism, two acid or base catalysts transfer the proton of O2 to O1 and the proton of C2 to C1 (Xu et al., 2016b; Rose, 1975). The catalytic mechanism of L-ribose synthesis from L-ribulose is shown in Figure 10.

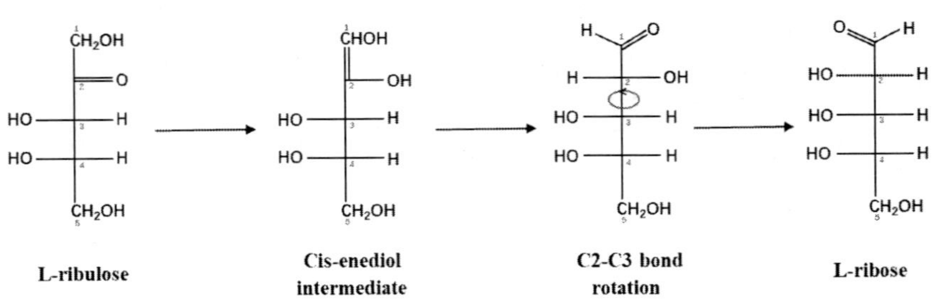

Figure 10. Catalytic mechanism of L-ribose synthesis from L-ribulose.

L-Ribose Biotransformation by using Isomerase

It was reported that the L-ribose isomerase (L-RI) from *Acinetobacter* sp. strain DL-28 was used for L-ribose biosynthesis with a conversion yield of 30% (Ahmed et al., 1999). A previous study was reported to introduce L-ribose isomerase activity into L-ribokinase-deficient *Escherichia coli* UP1110 and *Lactobacillus plantarum* BPT197 strains to catalyze the synthesis of L-ribose from L-arabinose in whole cells with the conversion rate of about 20% (Helanto et al., 2009). Yeo et al. constructed a yeast strain co-expressing L-AI and L-RI to convert L-arabinose to L-ribose by one-pot catalysis using a mixture of glucose and L-arabinose as a substrate and achieved a conversion yield of about 20% (w/w) (Yeo et al., 2021).

Mannose-6-phosphate isomerase (MPI) is also a key enzyme in the biosynthesis of L-ribose. Yeom et al. reported the production of L-ribose from L-ribulose using mannose-6-phosphate isomerase (MPI) from *Bacillus subtilis* with a 70% conversion yield (Yeom et al., 2009 a). A mixture of L-arabinose isomerase and mannose-6-phosphate isomerase purified from *Geobacillus thermodenitrificans* was used to synthesize L-ribose from L-arabinose in a one-pot method, the conversion yield was 23.6% (Yeom et al., 2009b). The L-arabinose isomerase and mannose 6-phosphate isomerase genes from *Geobacillus thermodenitrificans* were also co-expressed in recombinant *E. coli* for the whole-cell synthesis of L-ribose which achieved a conversion yield of about 33% (Kim, 2014). The highest conversion yield from L-arabinose to L-ribose reported so far is 35% (Li et al., 2021).

There are three main difficulties in synthesizing L-ribose by isomerases. First, the reaction substrate L-ribulose is not readily available. The main method is to synthesize L-ribulose using L-arabinose as a substrate, but the conversion yield is only about 10%. The reaction catalyzed by isomerase has the drawback of low conversion yield in general because of the thermodynamic limit of the enzyme, which decreases the production efficiency in industrial applications. Besides, the separation of L-ribose is difficult under low conversion yield (Lee et al., 2008).

Dehydrogenase for Biotransformation of L-Ribose

As shown in Figure 11, L-ribose can be synthesized from ribitol by using mannitol-1-dehydrogenase (MDH) from *Apium Graveolens,* which can overcome the thermodynamic equilibrium of isomerization reaction and have a relatively high conversion yield (Woodyer et al., 2008). The resting-cell catalytic method can be used, which allows the enzyme to maintain the location inside the original cells and the presence of the cell membrane makes it have higher stability, has the natural coenzyme regeneration system to regenerate NADH needed by MDH, and makes the process simple and of low cost by eliminating the enzyme extraction and purification steps.

Woodyer et al. reported that recombinant *E. coli* synthesizing L-ribose was constructed, and a conversion rate of 71% ± 4% at a ribitol concentration of 40 g/L was obtained under optimal conditions (Woodyer et al., 2008). In our unpublished study, the optimized recombinant *E. coli* resting cells catalyzed the synthesis of L-ribose with a conversion rate of 98% at a ribitol concentration of 40 g/L. This result is the highest among the reports, which is practical for the mass production of L-ribose. In summary, the advantages of using MDH to catalyze the synthesis of L-ribose are that it can achieve a very high conversion yield and there is almost no ribitol remained in the system.

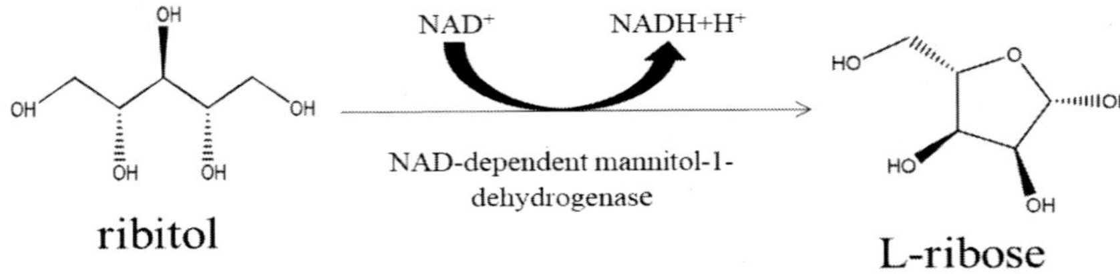

Figure 11. Single dehydrogenation reaction in the production of L-ribose from ribitol.

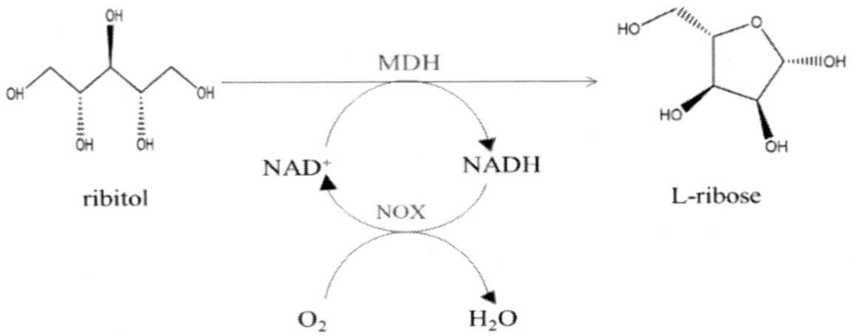

Figure 12. Schematic illustration of biocatalytic L-ribose production using *E. coli* whole cells harboring MDH coupled with NOX a cofactor regenerating enzyme.

The difficulty of MDH catalytic synthesis of L-ribose lies in the need for NADH when using high substrate concentration, which means that the content of NADH limits the reaction. To solve this problem, the expression of MDH and water-forming NADH oxidase (NOX) can be combined (Han and Eiteman, 2017; Tesfay et al., 2021). As shown in Figure 12, under the catalysis of MDH and NOX, the coenzyme regeneration cycle of NAD and NADH can be realized.

Secondly, the optimal temperature of MDH is 25°C, and the enzyme activity is completely lost at 40°C, which means its thermal stability is poor (Woodyer et al., 2008). Low thermal stability is not conducive to industrialization. Improving the thermostability can be achieved by directed evolution or rational design of the enzyme molecule. So far, only directed evolution has been reported to improve the thermal stability of MDH, and the mutant library was obtained by means of error-prone PCR. The conversion yield of 46.6% was obtained using MDH at 34 °C was, which has an overall increase of 19.2 times (Christ et al., 2010).

Finally, ribitol is currently an expensive sugar, which can be synthesized by hydrogenation reduction of D-ribose (Jeanloz et al., 1951). In summary, mannitol-1-dehydrogenase (MDH) as the key enzyme is a potential candidate for the industrial production of L-ribose if the above drawbacks are solved.

Conclusion

Rare sugars are a class of monosaccharides and their derivatives that exist in nature but with very little content, which have the characteristics of low calorie and low absorption. With the improvement of living standards, people's pursuit of healthy food and health preservation is also getting higher and higher, and traditional high-energy sweeteners such as sucrose tend to be replaced by low-calorie, functional sugars. In recent years, rare sugars have become a research hotspot.

There are many kinds of rare sugars, and different kinds of rare sugars have their own unique physiological functions and have huge application prospects in food, medicine, cosmetics, and other industries. Among them, D-allulose, D-allose, allitol, L-xylulose, L-xylose, and L-ribose are typical. In this chapter, the enzymes and methods for bioproduction of the above rare sugars were reviewed and discussed.

With the increasing attention to health, the market demand for rare sugars is also increasing, but their market size and application were limited by the high price. Therefore, it is extremely important to greatly reduce production costs and achieve large-scale production. Among them, finding or modifying a kind of enzyme catalyst with high catalytic activity and high heat resistance for industrial production of rare sugar is vital and is the main direction of the following efforts.

References

Aarnikunnas, J., Pihlajaniemi, A., Palva, A., Leisola, M. and Nyyssol, A. (2006). Cloning and expression of a xylitol-4-dehydrogenase gene from *Pantoea ananatis*. *Applied and Environmental Microbiology*, 368-377.

Alekseeva, A. A., Fedorchuk, V. V., Zarubina, S. A., Sadykhov, E. G., Matorin, A. D., Savin, S. S. and Tishkov, V. I. (2015). The role of Ala198 in the stability and coenzyme specificity of bacterial formate dehydrogenases. *Acta naturae*, 60-69.

Anderson, R. and Wood, W. (1962). Pathway of L-xylose and L-lyxose degradation in *Aerobacter aerogenes*. *Journal of Biological Chemistry*, 296-303.

Bai, W., Shen, J., Zhu, Y., Men, Y., Sun, Y. and Ma, Y. (2015), Characteristics and kinetic properties of L-rhamnose isomerase from *Bacillus subtilis* by isothermal titration calorimetry for the production of D-allose. *Food Science and Technology Research*, 13-22.

Bhuiyan, S. H., Ahmed, Z., Utamura, M. and Izumori, K. (1998). A new method for the production of L-lyxose from ribitol using microbial and enzymatic reactions. *Journal of Fermentation and Bioengineering*, 513-516.

Bhuiyan, S. H., Itami, Y., Izumori, K. (1997) Isolation of an l-Rhamnose Isomerase-Constitutive Mutant of *Pseudomonas sp.* Strain LL172: Purification and Characterization of the Enzyme. *Journal of Fermentation & Bioengineering* 84, 319-323.

Bicher, H. (1997). Antineoplastic compositions. *US Patent*, 624–908.

Chan, H. C., Zhu, Y., Hu, Y., Ko, T. P., Huang, C. H., Ren, F., Chen, C. C., Ma, Y., Guo, R. T. and Sun, Y. (2012). Crystal structures of D-psicose 3-epimerase from *Clostridium cellulolyticum* H10 and its complex with ketohexose sugars. *Protein Cell*, 123-131.

Chen, D., Chen, J., Liu, X., Guang, C., Zhang, W. and Mu, W. (2021). Biochemical identification of a hyperthermostable L-ribulose 3-epimerase from *Labedella endophytica* and its application for D allulose bioconversion. *International Journal of Biological Macromolecules*, 214-222.

Chen, J., Wu, H., Zhang, W. and Mu, W. (2020). Ribose-5-phosphate isomerases: characteristics, structural features, and applications. *Applied Microbiology and Biotechnology*, 6429-6441.

Chevrier, C., LeNouen, D., Defoin, A. and Tarnus, C. (2006). Synthesis of amino-L-lyxose phosphonates as fucosyl-phosphate mimics. *European Journal of Organic Chemistry*, 2384-2392.

Chow, C. M. and RajBhandary, U. L. (1993). Developmental regulation of the gene for formate dehydrogenase in *Neurospora crassa*. *Journal of bacteriology*, 3703-3709.

Christ, T. N., Deweese, K. A., and Woodyer, R. D. (2010). Directed evolution toward improved production of L-ribose from ribitol. *Combinatorial Chemistry & High Throughput Screening*, 302-308.

Davison, D. C. (1951). Studies on plant formic dehydrogenase. *The Biochemical journal*, 520-526.

Dimant, E. and Banay, M. (1960). Notes- preparation of L-xylose. *Journal of Organic Chemistry*, 475-476.

Doten, R. C. and Mortlock, R. P. (1985a). Characterization of xylitol-utilizing mutants of *Erwinia uredovora*. *Journal of Bacteriology*, 529-533.

Doten, R. C. and Mortlock R. P. (1985b). Production of D- and L-xylulose by mutants of *Klebsiella pneumoniae* and *Erwinia uredovora*. *Applied and Environmental Microbiology*, 158-162.

Doten, R. C., and Mortlock, R. P. (1985c). Inducible xylitol dehydrogenases in enteric bacteria. *Journal of Bacteriology*, 845-848.

El, K. O., Carbone, V. C., Ishikura, S. and Hara, A. (2010). Structure of the tetrameric form of human L-Xylulose reductase: probing the inhibitor-binding site with molecular modeling and site-directed mutagenesis. *Proteins*, 424–432.

Fang, Z., Zhang, W., Zhang, T., Guang, C. and Mu, W (2018). Isomerases and epimerases for biotransformation of pentoses. *Applied Microbiology and Biotechnology*, 7283-7292.

Frihed, T. G., Bols, M. and Pedersen, C. M. (2015). Synthesis of L-hexoses. *Chemical Reviews*, 3615-3676.

Gao, H., Kim, I. W., Choi, J. H., Khera, E., Wen, F. and Lee, J. K. (2015). Repeated production of l-xylulose by an immobilized whole-cell biocatalyst harboring l-arabinitol dehydrogenase coupled with an NAD^+ regeneration system. *Biochemical Engineering Journal*, 23-28.

Gao, H., Kim, T., Mardina, P., Zhou, P., Wen, F. and Lee, J. (2016). Rare sugar production by coupling of NADH oxidase and L-arabinitol dehydrogenase. *RSC Advances*, 66609-66616.

Granstrom, T. B., Takata, G., Morimoto, K., Leisola, M. and Izumori, K. (2005). L-Xylose and L-lyxose production from xylitol using *Alcaligenes* 701B strain and immobilized L-rhamnose isomerase enzyme. *Enzyme and Microbial Technology*, 976-981.

Hadipernata, M., Hayakawa, S. and Rachmat, R (2012). The utilization of rare sugars as a functional food. *In II Asia Pacific Symposium on Postharvest Research Education and Extension, APS2012* 1011, 369-374.

Han, Q. and Eiteman, M. A. (2017). Coupling xylitol dehydrogenase with NADH oxidase improves L-xylulose production in *Escherichia coli* culture. *Enzyme and Microbial Technology*, 106-113.

Han, W., Zhu, Y. and Bai, W. (2014a). Double-enzyme-coupled conversion of fructose to prepare mixed sugar solution containing rare sugars. *Chinese Journal of Biological Engineering*, 90-97.

Han, W., Zhu, Y., Men, Y., Yang, J., Liu, C. and Sun, Y. (2014b). Production of allitol from D-psicose by a novel isolated strain of *Klebsiella oxytoca* G4A4. *Journal of Basic Microbiology*, 1073-1079.

Hassanin, H. A., Eassa, M. A. and Jiang, B. (2018) Facile synthesis of bioactive allitol from D-psicose by coexpression of ribitol dehydrogenase and formate dehydrogenase in *Escherichia coli*. *Journal of Food Bioactives*, 117–122.

Hassanin, H. A., Letsididi, R., Koko, M. Y., Mu, W., Elferga, A. and Jiang, B. (2016a). Synthesis of allitol from D-psicose using ribitol dehydrogenase and formate dehydrogenase. *Tropical Journal of Pharmaceutical Research*, 2701-2708.

Hassanin, H. A., Mu, W., Koko, M. Y., Zhang, T., Masamba, K. and Jiang, B. (2016b). Allitol: production, properties and applications. *International Journal of Food Science and Technology*, 91-97.

Hassanin, H. A., Wang, X., Mu, W., Zhang, T. and Jiang, B. (2016c). Cloning and characterization of a new ribitol dehydrogenase from *Providencia alcalifaciens* RIMD 1656011. *Journal of the Science of Food and Agriculture*, 2917-2924.

Heikkilä, H., Koivikko, H., Ravanko, V., Alen, R., Nurmi, J., Ojamo, H., Haimi, P. and Tylli, M. (2002). Process for the production of xylose, *WO 02/088154*.

Heinz F., Hertel, S. and Vogel, M. (1998). Composition to reduce the level of sugar in the blood. *WO Patent*, 820-882.

Helanto, M., Kiviharju, K., Granström, T., Leisola, M., and Nyyssölä, A. (2009). Biotechnological production of L-ribose from L-arabinose. *Applied microbiology and biotechnology*, 77-83.

He, W., Jiang, B., Mu, W. and Zhang, T. (2016). Production of D-allulose with D-psicose 3-epimerase expressed and displayed on the surface of Bacillus subtilis spores. *Journal of Agricultural & Food Chemistry*, 7201.

Hiroshi, O., Seui, S., Hiroshi, S. and Toshitsugu, O. (1976). Increased urinary excretion of L-xylulose in patients with liver cirrhosis. *Clinica Chimica Acta*, 131-136.

Hofmann, C., Boll, R., Heitmann, B., Hauser, G., Durr, C., Frerich, A., Weitnauer, G., Glaser, S. J. and Bechthold, A. (2005). Genes encoding enzymes responsible for biosynthesis of L-lyxose and attachment of eurekanate during avilamycin biosynthesis. *Chemistry & Biology*, 1137-1143.

Hossain, M. A., Izuishi, K., Tokuda, M., Izumori, K. and Maeta, H. (2004) D-Allose has a strong suppressive effect against ischemia/reperfusion injury: a comparative study with allopurinol and superoxide dismutase. *Journal of Hepato-Biliary-Pancreatic Surgery*, 181-189.

Huang, Q., Xu, Z., Li, S., Xu, H., and Jiang, Y. (2019). Research progress in bio-manufacture pharmaceutical intermediate L-ribose. *Industrial microbiology (in Chinese)*, 61-69.

Huang, Z., Liu, M., Wang, B., Zhang, Y. and Cao, Z. (2006). Formate dehydrogenase and its application in cofactor NADH regeneration. *The Chinese Journal of Process Engineering*, 1011-1016.

Ishikura, S., Isaji, T., Usami, N., Kitahara, K., Nakagawa, J. and Hara, A. (2001). Molecular cloning, expression and tissue distribution of hamster diacetyl reductase. Identity with L-xylulose reductase. *Chemico-Biological Interactions*, 879-889.

Ishihara, Y., Katayama, K., Sakabe, M., Kitamura, M., Aizawa, M., Takara, M. and Itoh, K. (2011). Antioxidant properties of rare sugar D-allose: effects on mitochondrial reactive oxygen species production in neuro2A cells. *Journal of Bioscience & Bioengineering*, 638-642.

Itoh, H., Okaya, H., Khan, A. R., Tajima, S., Hayakawa, S. and Izumori, K. (1994). Purification and characterization of D-tagatose 3-epimerase from *Pseudomonas* sp. ST-24. *Bioscience Biotechnology and Biochemistry*, 2168-2171.

Itoh, H., Sato, T. and Izumori, K. (1995). Preparation of D-psicose from D-fructose by immobilized D-tagatose 3-epimerase. *Journal of Fermentation & Bioengineering*, 184-185.

Izumori, K. (2006). Izumoring: a strategy for bioproduction of all hexoses. *Journal of Biotechnology*, 717-722.

Izumori, K., Khan, A. R., Okaya, H. and Tsumura, T. (1993). A new enzyme, D-ketohexose 3-epimerase, from *Pseudomonas* sp. ST-24. *Bioscience Biotechnology and Biochemistry*, 1037-1039.

Jeanloz, R. W., and Fletcher Jr, H. G. (1951). The chemistry of ribose. *In Advances in Carbohydrate Chemistry*, 135-174.

Jokela, J., Pastinen, O. and Leisola, M., (2002). Isomerization of pentose and hexose sugars by an enzyme reactor packed with cross-linked xylose isomerase crystals. *Enzyme and Microbial Technology*, 67-76.

Khan, A. R., Tokunaga, H., Yoshida, K. and Izumori, K. (1991). Conversion of xylitol to L-xylulose by *Alcaligene* sp. 701B-cells. *Journal of Fermentation and Bioengineering*, 488-490.

Kim, K., Kim, H. J., Oh, D. K., Cha, S. S. and Rhee, S. (2006). Crystal structure of D-psicose 3-epimerase from *Agrobacterium tumefaciens* and its complex with true substrate D-fructose: a pivotal role of metal in catalysis, an active site for the non-phosphorylated substrate, and its conformational changes. *Journal of Molecular Biology*, 920-931.

Kim, K. R., Seo, E. S., and Oh, D. K. (2014). L-Ribose production from L-arabinose by immobilized recombinant *Escherichia coli* co-expressing the L-arabinose isomerase and mannose-6-phosphate isomerase genes from *Geobacillus thermodenitrificans*. *Applied biochemistry and biotechnology*, 275-288.

Kim, Y. S., Shin, K. C., Lim, Y. R. and Oh D. K. (2013). Characterization of a recombinant L-rhamnose isomerase from *Dictyoglomus turgidum* and its application for L-rhamnulose production. *Biotechnology Letters*, 259-264.

Labrou, N. E., Rigden, D. J. (2001) Active-site characterization of *Candida boidinii* formate dehydrogenase. *The Biochemical journal*, 455-463.

Leang, K., Takada, G., Fukai, Y., Morimoto, K., Granstrom, T. B. and Izumori, K. (2004a). Novel reactions of L-rhamnose isomerase from *Pseudomonas stutzeri* and its relation with D-xylose isomerase via substrate specificity. *Biochimica et Biophysica Acta (BBA)-General Subjects*, 68-77.

Leang, K., Takada, G., Ishimura, A., Okita, M. and Izumori, K. (2004b). Cloning, nucleotide sequence, and overexpression of the L-rhamnose isomerase gene from *Pseudomonas stutzeri* in *Escherichia coli*. *Applied and Environmental Microbiology*, 3298-3304.

Lee, S. H., Lee, E., and Kim, I. H. (2008). Simulation of SMB [simulated moving bed] chromatography for separation of L-ribose and L-arabinose by ASPEN chromatography. *KSBB Journal*, 135-141.

Li, M., Zhu, W., Meng, Q., Miao, M. and Zhang, T., (2021). Characterization of xylitol 4-dehydrogenase from *Erwinia aphidicola* and its co-expression with NADH oxidase in *Bacillus subtilis*. *Process Biochemistry*, 92-100.

Lin, C. J., Tseng, W. C. and Fang, T. Y. (2011). Characterization of a thermophilic L-rhamnose isomerase from *Caldicellulosiruptor saccharolyticus* ATCC 43494. *Journal of Agricultural and Food Chemistry*, 8702-8708.

Lin, C. J., Tseng, W. C., Lin, T. H., Liu, S. M., Tzou, W. S. and Fang, T. Y. (2010). Characterization of a thermophilic L-rhamnose isomerase from *Thermoanaerobacterium saccharolyticum* NTOU1. *Journal of Agricultural and Food Chemistry*, 10431-10436.

Lin, J., Chai, Y., Ren Y., Sun, X., Wen, X. and Lin, J. (2020a). A triangular reticular RNA scaffold for intracellular fixation of recombinant proteins and its construction method and application. CN111378654A [P].

Lin, J., Chai, Y., Wen X., Ren, Y., Sun, X., Song, X. and Lin, J. (2020b). Method of recombinant enzyme was immobilized on a recombinant intracellular RNA scaffold to promote the production of allitol. China, CN111378612A [P].

Li, Z., Gao, Y., Nakanishi, H., Gao, X. and Cai, L. (2013). Biosynthesis of rare hexoses using microorganisms and related enzymes. *Beilstein Journal of Organic Chemistry*, 2434-2445.

Li, Z., Li, Y., Duan, S., Liu, J., Yuan, P., Nakanishi, H. and Gao, X. D. (2015). Bioconversion of D-glucose to D-psicose with immobilized D-xylose isomerase and D-psicose 3-epimerase on *Saccharomyces cerevisiae* spores. *Journal of Industrial Microbiology & Biotechnology*, 1117-1128.

Li, Z., Liu, X., Nakanishi, H., and Gao, X. D. (2020). Encapsulation of mannose-6-phosphate isomerase in *Yeast Spores* and its application in L-ribose production. *Journal of Agricultural and Food Chemistry*, 6892-6899.

London, J. (1990). Uncommon pathways of metabolism among lactic acid bacteria. *FEMS Microbiology Reviews*, 103-112.

Lu, F., Xu, W., Zhang, W., Guang, C. and Mu, W. (2019) Polyol dehydrogenases: intermediate role in the bioconversion of rare sugars and alcohols. *Applied Microbiology and Biotechnology*, 6473-6481.

Mahmud, A., Hattori, K., Hongwen, C., Kitamoto, N., Suzuki, T. and Nakamura, K. (2013). Xylitol production by NAD^+-dependent xylitol dehydrogenase (xdhA)- and L-arabitol-4-dehydrogenase (ladA)-disrupted mutants of *Aspergillus oryzae*. *Journal of Bioscience and Bioengineering*, 353-359.

Mathé C., Gosselin, G. (2006). L-Nucleoside enantiomers as antivirals drugs: a mini-review. *Antiviral Research*, 276-281.

Menavuvu, B. T., Poonperm, W., Takeda, K., Morimoto, K., Granström, T. B., Takada, G., and Izumori, K. (2006). Novel substrate specificity of D-arabinose isomerase from *Klebsiella pneumoniae* and its application to production of D-altrose from D-psicose. *Journal of Bioscience and Bioengineering*, 436-41.

Meng, Q., Zhang, T., Jiang, B., Mu, W. and Miao, M. (2016). Advances in applications, metabolism, and biotechnological production of L-xylulose. *Applied Microbiology and Biotechnology*, 535-540.

Min, J. S., Choi, J. H., Kang, S. H., Shin, K. C. and Oh, D. K. (2017). Characterization of L-rhamnose isomerase from *Clostridium stercorarium* and its application to the production of D-allose from D-allulose (D-psicose). *Biotechnology Letters*, 325–334.

Muniruzzaman, S., Kunihisa, Y., Ichiraku, K. and Izumori, K. (1995a) Purification and characterization of a ribitol dehydrogenase from *Enterbacter agglomerans* strain 221e. *Journal of Fermentation and Bioengineering*, 496-498.

Muniruzzaman, S., Pan, Y. T., Zeng, Y., Atkins, B., Izumori, K. and Elbein, A. D. (1996). Inhibition of glycoprotein processing by L-fructose and L-xylulose. *Glycobiology*, 795-803.

Muniruzzaman, S., Tokunaga, H. and Izumori, K. (1995b). Conversion of D-psicose to allitol by *Enterobacter agglomerans* strain 221e. *Journal of Fermentation and Bioengineering*, 323-327.

Nanba, H., Takaoka, Y. and Hasegawa, J. (2003). Purification and characterization of formate dehydrogenase from *Ancylobacter aquaticus* strain KNK607M, and cloning of the gene. *Bioscience, Biotechnology, and Biochemistry*, 720-728.

Oosaka, K. (2009). Possible as monosaccharide laxative of rare sugar alcohols. *Yakugaku Zasshi*, 575-580.

Ottens, M., Lebreton, B., Zomerdijk, M., Rijkers, M. P. W. M., Bruinsma, O. S. L., and van der Wielen, L. A. M. (2004). Impurity effects on the crystallization kinetics of ampicillin. *Industrial & Engineering Chemistry Research*, 7932-7938.

Park, C. S. (2014). Characterization of a recombinant L-rhamnose isomerase from *Bacillus subtilis* and its application on production of L-lyxose and L-mannose. *Biotechnol*ogy and *Bioprocess Eng*ineering, 18-25.

Park, C. S., Yeom, S. J., Lim, Y. R., Kim, Y. S. and Oh, D. K. (2010). Characterization of a recombinant thermostable L: -rhamnose isomerase from *Thermotoga maritima* ATCC 43589 and its application in the production of L-lyxose and L-mannose. *Biotechnology Letters*, 1947-1953.

Pierce, S. B., Spurrell, C. H., Mandell, J. B., Lee, M. K., Zeligson, S., Bereman, M. S., Stray, S. M., Fokstuen, S., MacCoss M. J. and Levy-Lahad, E. (2011). Garrod's fourth inborn error of metabolism solved by the identification of mutations causing pentosuria. *Proceedings of the National Academy of Sciences of the United States of America*, 18313-18317.

Pometun, A. A., Boyko, K. M., Yurchenko, T. S., Nikolaeva, A. Y., Kargov, I. S., Atroshenko, D. L., Savin, S. S., Popov, V. O. and Tishkov, V. I. (2020). Highly-active recombinant formate dehydrogenase from pathogenic bacterium *Staphylococcus aureus*: preparation and crystallization. *Biochemistry (Moscow)*, 689-696.

Poonperm, W., Takata, G., Morimoto, K., Granström, T. B. and Izumori, K. (2007a). Production of L-xylulose from xylitol by a newly isolated strain of *Bacillus pallidus* Y25 and characterization of its relevant enzyme xylitol dehydrogenase. *Enzyme and Microbial Technology*, 1206-1212.

Poonperm, W., Takata, G., Okada, H., Morimoto, K., Granstrom, T. B. and Izumori, K. (2007b). Cloning, sequencing, overexpression and characterization of L-rhamnose isomerase from *Bacillus pallidus* Y25 for rare sugar production. *Applied Microbiology and Biotechnology*, 1297-1307.

Prabhu, P., Thanh, T., Ngoc, D., Jeya, M., Kang, L. W. and Lee, J. K. (2011). Cloning and characterization of a rhamnose isomerase from *Bacillus halodurans*. *Applied Microbiology and Biotechnology*, 635-644.

Richard, P., Londesborough, J., Putkonen, M., Kalkkinen, N. and Penttila, M. (2001). Cloning and expression of a fungal L-arabinitol 4-dehydrogenase gene. *Journal of Biological Chemistry*, 40631-40637.

Rojkova, A. M., Galkin, A. G., Kulakova, L. B., Serov, A. E., Savitsky, P. A., Fedorchuk V. V. and Tishkov, V. I. (1999). Bacterial formate dehydrogenase. Increasing the enzyme thermal stabilityby hydrophobization of alpha-helices. *FEBS Letters*, 183-188.

Rose, I. A. (1975). Mechanism of the aldose-ketose isomerase reactions. *Advances in Enzymology and Related Areas of Molecular Biology*, 491-517.

Sagurthi, S. R., Gowda, G., Savithri, H. S., and Murthy, M. R. N. (2009). Structures of mannose-6-phosphate isomerase from *Salmonella typhimurium* bound to metal atoms and substrate: implications for catalytic mechanism. *Acta Crystallographica Section D: Biological Crystallography*, 724-732.

Shimonishi, T., and Izumori, K. (1996). A new enzyme, L-ribose isomerase from *Acinetobacter sp.* strain DL-28. *Journal of Fermentation and Bioengineering,* 493-497.

Shin, S. M., Cao, T. P., Choi, J. M., Kim, S. B., Lee, S. J., Lee, S. H. and Lee, D. W. (2017). TM0416, a hyperthermophilic promiscuous nonphosphorylated sugar isomerase, catalyzes various C-5 and C-6 epimerization reactions. *Applied and Environmental Microbiology*, 1-16.

Singh, R. K., Tiwari, M. K., Singh, R., Haw, J. R. and Lee, J. K. (2014). Immobilization of L-arabinitol dehydrogenase on aldehyde-functionalized silicon oxide nanoparticles for L-xylulose production. *Applied Microbiology and Biotechnology*, 1095–1104.

Sui, L., Dong, Y., Watanabe, Y., Yamaguchi, F., Hatano, N., Tsukamoto, I., Izumori, K. and Tokuda, M. (2005). The inhibitory effect and possible mechanisms of D-allose on cancer cell proliferation. *International Journal of Oncology*, 907-912.

Takata, G., Poonperm, W., Morimoto, K. and Izumori, K. (2010). Cloning and overexpression of the xylitol dehydrogenase gene from *Bacillus pallidus* and its application to L-xylulose production. *Bioscience Biotechnology and Biochemistry*, 1807-1813.

Takata, G., Uechi, K., Taniguchi, E., Kanbara, Y., Yoshihara, A., Morimoto, K. and Izumori K. (2011). Characterization of *Mesorhizobium loti* L-rhamnose isomerase and its application to L-talose production. *Bioscience Biotechnology and Biochemistry*, 1006-1009.

Takeshita, K., Ishida, Y., Takata, G. and Izumori, K. (2000) Direct production of allitol from D-fructose by a coupling reaction using D-tagatose-3-epimerase, ribitol dehydrogenase and formate dehydrogenase. *Journal of Bioscience and Bioengineering*, 545-548.

Tesfay, M. A., Wen, X., Lin, H., Liu, Y., Li, C., Chen, L., Lin, J, and Lin, J. (2021a). Efficient L-xylulose production using whole-cell biocatalyst with NAD^+ regeneration system through co-expression of xylitol dehydrogenase and NADH oxidase in *Escherichia coli*. *Biochemical Engineering Journal*, 108137.

Tesfay, M. A., Wen, X., Liu, Y., Lin, H., Chen, L., Lin, J. and Lin, J., (2021b). Construction of recombinant *Escherichia coli* expressing xylitol-4-dehydrogenase and optimization for enhanced L-xylulose biotransformation from xylitol. *Bioprocess and Biosystems Engineering*, 1021-1032.

Tishkov, V. I. and Popov, V. O. (2004). Catalytic mechanism and application of formate dehydrogenase. *Biochemistry (Moscow)*, 1252-1267.

Tiwari, M. K., Singh, R. K., Gao, H., Kim, T., Chang, S. and Kim, H. S. (2014). pH-rate profiles of L-arabinitol 4-dehydrogenase from *Hypocrea jecorina* and its application in L-xylulose production. *Bioorganic & Medicinal Chemistry Letters*, 173-176.

Tseng, C. W., Liao, C. Y., Sun, Y., Peng, C. C., Tzen, J. T., Guo, R. T. and Liu, J. R. (2014). Immobilization of *Clostridium cellulolyticum* D-psicose 3-epimerase on artificial oil bodies. *Journal of Agricultural & Food Chemistry*, 6771-6776.

Uechi, K., Takata, G., Fukai, Y., Yoshihara, A., and Morimoto, K. (2013). Gene cloning and characterization of L-ribulose 3-epimerase from *Mesorhizobium loti* and its application to rare sugar production. *Bioscience Biotechnology and Biochemistry*, 511-515.

Usvalampi, A., Kiviharju, K., Leisola, M. and Nyyssola, A. (2009). Factors affecting the production of L-xylulose by resting cells of recombinant *Escherichia coli*. *Journal of Industrial Microbiology & Biotechnology*, 1323-2133.

Usvalampi, A., Turunen, O., Valjakka, J., Pastinen, O., Leisola, M. and Nyyssola, A. (2012). Production of L-xylose from L-xylulose using *Escherichia coli* L-fucose isomerase. *Enzyme and Microbial Technology*, 71-76.

Wang, P., Hong, J. H., Cooperwood, J. S., and Chu, C. K. (1998). Recent advances in L-nucleosides: *chemistry and biology*. *Antiviral Research*, 19-44.

Wen, P., Wu, D., Zheng, P., Chen, P., Liu, S. and Fu, Y. (2019). Highly efficient biosynthesis of heliotropin by engineered *Escherichia coli* coexpressing trans-anethole oxygenase and formate dehydrogenase. *Journal of Agricultural and Food Chemistry*, 14121-14128.

Wen, X., Lin, H., Ren, Y., Li, C., Zhang, C., Lin, J. and Lin, J. (2022). Allitol bioproduction by recombinant *Escherichia coli* with NADH regeneration system co-expressing ribitol dehydrogenase (RDH) and formate dehydrogenase (FDH) in individual or in fusion. *Electronic Journal of Biotechnology*, 91-98.

Wen, X., Lin, H., Ren, Y., Li, C., Zhang, C., Lin, J. and Lin, J. (2020a). Efficient allitol bioproduction from D-fructose catalyzed by recombinant *E. coli* whole cells, and the condition optimization, product purification. *Applied Biochemistry and Biotechnology*, 680-697.

Wen, X., Lin, H., Ren, Y., Li, C., Zhang, C., Lin, J. and Lin, J. (2020b) Optimization for allitol production from D-glucose by using immobilized glucose isomerase and recombinant *E. coli* expressing D-psicose-3-epimerase, ribitol dehydrogenase and formate dehydrogenase. *Biotechnology Letters*, 2135-2145.

Woodyer, R. D., Wymer, N. J., Racine, F. M., Khan, S. N., and Saha, B. C. (2008). Efficient production of L-ribose with a recombinant *Escherichia coli* biocatalyst. *Applied and Environmental Microbiology*, 2967-2975.

Xu, H., Liu, C., Xu, Z., Wang, X., Li, S., and Feng, X. H. (2016a). An L-ribose isomerase and its application in bio-production of L-ribose. Chinese patent no. 201610218779.3.

Xu, Z., Sha, Y., Liu, C., Li, S., Liang, J., Zhou, J., & Xu, H. (2016b). L-ribose isomerase and mannose-6-phosphate isomerase: properties and applications for L-ribose production. *Applied microbiology and biotechnology*, 9003-9011.

Xu, Z., Wang, R., Liu, C., Chi, B., Gao, J., Chen, B., and Xu, H. (2016c). A new L-arabinose isomerase with copper ion tolerance is suitable for creating protein–inorganic hybrid nanoflowers with enhanced enzyme activity and stability. *RSC advances*, 30791-30794.

Yamaguchi, F., Kamitori, K. and Sanada, K. (2008) Rare sugar D-allose enhances anti-tumor effect of 5-fluorouracil on the human hepatocellular carcinoma cell line HuH-7. *Journal of Bioscience & Bioengineering*, 248-252.

Yamasaki-Yashiki, S., Komeda, H., Hoshino. K. and Asano, Y. (2017). Characterization and gene cloning of L-xylulose reductase involved in L-arabinose catabolism from the pentose-fermenting fungus *Rhizomucor pusillus*. *Bioscience Biotechnology and Biochemistry*, 81-87.

Yang, W. B., Patil, S. S., Tsai, C. H., Lin, C. H. and Fang, J. M. (2002). The synthesis of L-gulose and L-xylose from D-gluconolactone. *Tetrahedron*, 253-259.

Yeo, I. S., Cho, B. K., and Kim, J. H. (2021). Conversion of L-arabinose to L-ribose by genetically engineered *Candida tropicalis*. *Bioprocess and biosystems engineering*, 1147-1154.

Yeom, S. J., Ji, J. H., Kim, N. H., Park, C. S., and Oh, D. K. (2009a). Substrate specificity of a mannose-6-phosphate isomerase from *Bacillus subtilis* and its application in the production of L-ribose. *Applied and environmental microbiology*, 4705-4710.

Yeom, S. J., Kim, N. H., Park, C. S., and Oh, D. K. (2009b). L-Ribose production from L-arabinose by using purified L-arabinose isomerase and mannose-6-phosphate isomerase from *Geobacillus thermodenitrificans*. *Applied and Environmental Microbiology*, 6941-6943.

Yoshida, H., Wayoon, P., Takada, G., Izumori, K. and Kamitori, S. (2006). Crystallization and preliminary X-ray diffraction studies of L-rhamnose isomerase from *Pseudomonas stutzeri*. *Acta Crystallographica Section F-Structural Biology Communications*, 550-552.

Yoshida, H., Yamada, M., Nishitani, T., Takada, G., Izumori, K. and Kamitori, S. (2007). Crystal structures of D-tagatose 3-epimerase from *Pseudomonas cichorii* and its complexes with D-tagatose and D-fructose. *Journal of Molecular Biology*, 443-453.

Yoshida, H., Yoshihara, A., Kato, S., Mochizuki, S., Akimitsu, K., Izumori, K. and Kamitori, S. (2021). Crystal structure of a novel homodimeric L-ribulose 3-epimerase from *Methylomonus* sp. *FEBS Open Bio*, 1621-1637.

Yoshida, H., Yoshihara, A., Teraoka, M., Terami, Y., Takata, G., Izumori, K., and Kamitori, S. (2014). X-ray structure of a novel L-ribose isomerase acting on a non-natural sugar L-ribose as its ideal substrate. *The FEBS Journal*, 3150-3164.

Yoshihara, A., Kozakai, T., Shintani, T., Matsutani, R., Ohtani, K., Iida, T., Akimitsu, K., Izumori, K. and Gullapai, P. K. (2017). Purification and characterization of D-allulose 3-epimerase derived from *Arthrobacter globiformis* M30, a GRAS microorganism. *Journal of Bioscience and Bioengineering*, 170-176.

Zhang, W., Fang, D., Xing, Q., Zhou, L., Jiang, B. and Mu, W. (2013). Characterization of a novel metal-dependent D-psicose 3-epimerase from *Clostridium scindens* 35704. *Plos One*, e62987.

Zhang, X., Zhu, M., Han, R., Zhao, Y., Chen, K., Qian, K., Shao, M., Yang, T., Xu, M., Xu, J. and Rao, Z. (2019) A novel 3-phytosterone-9α-hydroxylase oxygenation component and its application in bioconversion of 4-androstene-3,17-dione to 9α-hydroxy-4-androstene-3,17-dione coupling with a NADH regeneration formate dehydrogenase. *Molecules*, 2534.

Zhang, Y., Zhao, X., Yang, L., Han, Y. and Liu, J. (2014). Isolation and identification of a novel L-xylulose-producing strain. *Food Sci.*, 99-203.

Zhu, C. Y., Zhu, Y. H., Zhou, H. P., Xu, Y. Y., Gao, J. and Zhang, Y. W. (2021). Cloning, expression, and characterization of an arabitol dehydrogenase and coupled with NADH oxidase for effective production of L-xylulose. *Preparative Biochemistry & Biotechnology*, 1-8.

Zhu, Y., Li, H., Liu, P., Yang, J., Zhang, X. and Sun, Y. (2015). Construction of allitol synthesis pathway by multi-enzyme coexpression in *Escherichia coli* and its application in allitol production. *Journal of Industrial Microbiology & Biotechnology*, 661-669.

Chapter 2

Functional and Bioenergetic Effects of ATP-Sensitive Potassium Transport in Mitochondria

Olga V. Akopova*

Circulation Department, Bogomoletz Institute of Physiology,
NAS of Ukraine, Kiev, Ukraine

Abstract

ATP-sensitive potassium uptake conducted by mitochondrial ATP-sensitive potassium channels (mKATP channels) is presently recognized as a powerful tool of tissues protection under several pathophysiological states and conditions. Cytoprotective signaling triggered by mKATP channels opening is largely based on the modulation of mitochondrial bioenergetics. ATP-sensitive K^+ transport is an all-round modulator of mitochondrial functions and energy-dependent processes: potassium cycle, oxygen consumption, ATP synthesis, Ca^{2+} uptake and ROS production. The modulation of these processes is based on the abundance of mKATP channels in mitochondrial membrane, on the rate of potassium transport, and the contribution of ATP-sensitive K^+ transport to oxygen consumption in mitochondria. For understanding the impact of mKATP channels opening on mitochondrial bioenergetics, it is important to know how functional effects of ATP-sensitive K^+ transport (the rate of K^+ uptake and its share in oxygen consumption) are transformed into bioenergetic ones: the modulation of membrane potential, uncoupling of the respiratory chain and related energy-dependent processes (Ca^{2+} transport, ATP synthesis). It is strangely enough, but there is still no consensus in the literature about the direct effects of ATP-sensitive K^+ transport on mitochondrial functions. The use of pharmacological

* Corresponding Author's Email: olga.akopova01@mail.ru.

In: Advances in Biology. Volume 1
Editor: Charles D. Grant
ISBN: 979-8-88697-201-6

modulators is the most common way to assess mKATP channels functioning in mitochondria. However, one crucial point of these studies is the lack of specificity of commonly used pharmacological modulators of mKATP channels, the openers (diazoxide, pinacidil, nicorandil, chromacalim), and the blockers (glibenclamide and 5-hydroxydecanoate). In this chapter the properties of mKATP channels, functional and bioenergetic effects of ATP-sensitive K^+ transport are analyzed based on the authors' studies on isolated mitochondria. Also, the literary data on the properties of mKATP channel, cellular mechanisms of tissue protection, bioenergetic effects of mKATP channels opening, and the prospectives in mKATP channels studies are discussed.

Keywords: mitochondria, mKATP channels, potassium cycle, oxygen consumption, bioenergetics, cytoprotection

Abbreviations:

mKATP channel – mitochondrial ATP-sensitive K+ channel;
DZ – diazoxide;
PKCε – protein kinase C epsilon;
PKG – protein kinase G;
HIF – hypoxia inducible factor;
Kir6.1, Kir6.2 – potassium inward rectifier 6.1 (6.2) subunit;
SUR – sulfonylurea receptor;
KCO – K+ channels openers;
$\Delta\Psi_m$ – mitochondrial membrane potential;
mPTP – permeability transition pore

Introduction

Cytoprotective Effects of MKATP Channels Opening: Functional and Bioenergetic Aspects

ATP-sensitive potassium uptake conducted by mitochondrial ATP-sensitive potassium channels (mKATP channels) is presently recognized as a powerful tool of tissues protection under several pathophysiological states and conditions [1-4]. According to contemporary knowledge, mechanisms of cytoprotection are based on cytoprotective signaling to mKATP channels, remodeling of mitochondrial functions and bioenergetics by mKATP channels opening, and triggering downstream signaling events, which putative primary target is glycogen synthase kinase-3β (GSK3β and permeability transition pore (mPTP), a gateway to cellular apoptosis and necrosis [1, 5-7]. This concept was developed in pioneering works of Garlids' group [1] and was not much changed over past decade [2].

MKATP channels opening has been involved in multiple cell and tissues protection effects under several diseases and pathophysiological conditions [4, 8]. Protective mechanisms included the improvement of mitochondrial functions and dynamics [9, 10], improvement of mitochondrial energy state and ATP synthesis [11], prevention of the development of cardiac and neuronal dysfunction [12-14]. In ischemia/reperfusion (I/R) models mKATP channels opening reduced I/R injury and produced anti-apoptotic, anti-inflammatory, and anti-oxidant effects, as shown in heart, brain and liver tissues [15-19]. Adaptive responses to hypoxia as well included activation [20, 21] or up-regulation of mKATP channel [22], and the channel involvement in the activation of HIF-dependent pathways [23, 24].

For decades, the narrow place in the understanding of the mechanisms of tissues protection by mKATP channels opening was the absence of data on molecular basis of ATP-sensitive K^+ transport, thus poor knowledge of protein interaction network involved in the realization of cytoprotective effects on molecular level. Recent discoveries shed more light on the molecular composition of mKATP channel, but put still more questions regarding the involvement of mKATP channels in cellular mechanisms of tissues protection, their molecular basis and physiological relevance.

Bioenergetic effects of mKATP channels opening are one of the key issues of cytoprotective mechanisms afforded by pharmacological mKATP channels openers [1-3]. However, there is still much diversity in the reported effects of mKATP channels opening on mitochondrial energy state, ROS production, Ca^{2+} transport, and ATP synthesis. In addition, an understanding of the role of mKATP channels in the mechanisms of tissue protection requires better knowledge of functional effects of ATP sensitive K^+ transport *in vivo*, but still there is scarce evidence about functional consequences of mKATP channels opening in a living organism. So, an understanding of physiological relevance of bioenergetics effects of mKATP channels opening needs an outline of cellular mechanisms involved in the regulation of mKATP channels activity, a brief overview of the current concepts of the molecular nature and properties of mKATP channel, and the discussion of the direct bioenergetics effects of ATP-sensitive K^+ transport and their functional consequences in mitochondria.

Cytoprotective Signaling to and from MKATP Channel

Cardio- and neuroprotective effects of mKATP channels opening were shown in multiple studies [1-4, 8]. Pharmacological mKATP channels openers exhibited high efficiency against ishemic and hypoxic conditions, and the action of cytotoxic agents [6, 23-27]. However, main limitation in the application of mKATP openers in clinics was their poor selectivity towards different isoforms of mKATP channel and multiple off-target effects [3]. This largely prevented the progress in the understanding of molecular mechanisms of cytoprotection afforded by these drugs.

Most detailed information about the mechanisms of cardio- and neuroprotective effects afforded by mKATP opening came from the studies of so-called ishemic pre- and post-conditioning (IPC). Not only the involvement of native mKATP channels in these phenomena was shown, but also the ability of pharmacological openers of mKATP channels (diazoxide, pinacidil, nicorandil, etc.) to effectively mimic the effects of IPC *via* mKATP channels opening [1, 6-8, 19, 23].

IPC is a receptor-mediated phenomenon [1, 28, 29], which starting point is the binding of physiologically active agents released under ischemia (adenosine, bradykinine, endothelin, opioid peptides, and others) respectively to their G_i proteins coupled receptors (G_iPCR), and triggering protective signaling pathways: the activation of phosphatidylinositol 3-kinase (PI3K), release of inositol-triphosphate and the activation of downstream phospholipid-dependent kinases (PDK) and Akt (proteinkinase B) [1, 7, 30]. Akt is capable of activating constitutive NO synthase and trigger NO/cGMP/PKG-mediated signaling [1, 28]. This survival signaling eventually results in PKC activation [1, 28], which is mediated by mKATP channels and ROS.

Generally, ishemic and hypoxic conditions are accompanied by ROS overproduction, which main sources are NADPH oxidase (NOX) and the respiratory chain [31]. MKATP channel can be activated by ROS either directly [1-3], or *via* upstream ROS/RNS-dependent signaling pathways mediated by PI3K/Akt, NO/cGMP/PKG, and PKCε [30-33]. MKATP channel opening in turn could trigger ROS-dependent signaling by the elevation of ROS production in mitochondria. Thus, activation of PKCε, Akt, p38MAPK,

$ERK_{1/2}$ downstream of mKATP channels opening was shown [19, 26, 30, 34, 35]. These literary data indicated the ability of mKATP channel to accept signals from cytosol to mitochondria and convey signaling from mitochondria to cytosol as a feedback mechanism [31], which is in line with the concept of mKATP channel as a 'ROS sensor' [2], capable of attenuation of ROS overproduction [2, 17, 26], or, on the contrary, triggering of ROS-mediated signaling [1, 31], dependent on the direct effect of mKATP channel opening on ROS production in mitochondria. PKG and PKCε are most proximal physiological activators of mKATP channel [1-3, 32-34]. Key role in PKCε-mediated signaling to and from mKATP channels belongs to the modulation of ROS production. In several studies ROS-dependent PKCε activation upstream and downstream of mKATP channel was shown [1, 2, 31, 34].

It is well known that PKC is activated by hydroperoxide, and is capable of direct activation of KATP channel [36, 37]. Direct activation of mKATP channel by PKCε and cGMP/PKG was shown on isolated heart mitochondria [38]. The above studies proved the existence of molecular basis for upstream mKATP channel activation by PKCε; however detailed molecular mechanism remains elusive. First, mKATP channels opening and resulting elevation of H_2O_2 production could trigger downstream signaling by the activation of mitochondially localized PKCε [1, 31]. This mechanism could directly link mKATP opening to mPTP closure by the inhibition of GSK3β [1, 7]. MKATP channels opening could trigger downstream signaling mediated by mitochondrial and cytosolic PKCε [31, 34], MAPK, and ERK1/2 kinases, highly sensitive to free radicals production [18, 19, 35]. MAPK/ERK-signaling downstream of mKATP opening protected the heart under IPC [19, 39]. The activation of MAPK/ERK pathway correlated with overexpression of mKATP channel in malignant glioma cells, which could explain the radioresistance of this cell type [35].

So, participation of mKATP channel in cytoprotective signaling largely was dependent on the direct effect of mKATP channels opening on ROS production. While the elevation of ROS production by mKATP opening triggered cytoprotective ROS-dependent signaling pathways, the reduction of ROS production caused by the mild uncoupling of the respiratory chain by ATP-sensitive K^+ transport was capable to reduce ROS and RNS overproduction and block ROS-dependent cell death pathways. As it was shown, protection of neurons against glutamate-, methylmalonate, and 3-nitropropionic acid cytotoxicity was mediated by the suppression of ROS and RNS production by mKATP channels openers [25, 40, 41]. In the heart, suppression of ROS production by mKATP channels opening protected against doxorubicin cytotoxicity [42], blocked cell death pathways, reduced tissue damage caused by I/R [7, 15, 17, 30], and protected cardiomyocytes from hypoxic injury by activating HIF-dependent pathways [23, 24]. Anti-inflammatory effect of mKATP channel opener diazoxide was shown in heart, brain, and liver [15, 16, 19].

The diversity of the direct effects of mKATP opening on ROS production in mitochondria still remains an unresolved issue in mitochondrial bioenergetics. Different energy state of the sites of ROS formation in the respiratory chain, dependent on experimental conditions and mitochondrial preparations, could be a plausible explanation of such diversity of the effects. As it was shown, diazoxide was capable both to increase and to reduce ROS production in heart mitochondria [43]. Thus, it was shown that diazoxide reduced ROS production in the course of forward electron transport, but increased ROS produced by reverse electron transport (RET), which was blocked by rotenone [43]. Because of mild uncoupling caused by ATP-sensitive K^+ transport, the electron flow to the sites with highly negative electrochemical potential was rather unlikely. But taking into account that a potential of -160 mV is required for superoxide formation [44], we suppose that a 'limited RET' to the less negatively charged sites of Complex I still was possible, which could allow for the elevation of ROS production by diazoxide. Other mKATP channels opener nicorandil too increased ROS production, but at Complex III, which effect was blocked by 5-HD [45]. Generally, experiments *in situ* (e.g., on isolated myocardium) do not allow for controlling such conditions, which in our opinion could explain the diversity of data obtained in different laboratories.

One crucial issue in understanding the mechanisms of cytoprotective effects afforded by mKATP channels opening is the lack of knowledge about proteins involved in the interaction with mKATP channel. So, it was shown that under hypoxia and ishemia cardioprotective effects of DZ were mediated by the activation of HIF-1/HRE pathway [23, 24]. Protection of cardiomyocytes under hypoxia too was mediated by the interaction of Kir6.2 subunit with heat shock protein 90 (Hsp-90) [46]. Also, the interaction of Kir6.1 subunit with gap junction protein connexin43 and PKCε was required for cardioprotective effects of mKATP opening [47]. Protection was abolished by the silencing or pharmacological inhibition of Hsp-90 or connexin43 [46, 47]. Cardioprotective effect of DZ too required the activity of mitochondrial connexin43 [27]. However, these data are insufficient to reveal the protein network involved in signaling pathways mediated by mKATP channel.

Currently, further progress in understanding of molecular mechanisms of cytoprotection afforded by mKATP opening is largely slowed down by apparent multiplicity of ATP-sensitive K^+ channels found in mitochondria [48]. So, in the light of earlier research [49, 50] and recent discoveries [51, 52], it could be suggested that mitochondria may possess several ATP-sensitive K^+ channel types including proteins capable to exhibit similar K^+ conductant properties under specific conditions, as it was shown for F_0F_1 ATP synthase [52]. Physiological relevance of these channels is a critical issue in cardio- and neuroprotection. It is reasonable to hypothesize that different mKATP channels types could be involved in specific to these types receptor-mediated pathways, also dependent on cell types. This altogether makes the study of the involvement of mKATP channel in the mechanisms of cytoprotection a very complicated task, which requires the knowledge of molecular basis of signaling pathways and cannot be resolved without highly elaborated genetic and pharmacological tools. In view of the above concerns, we need a brief overview of the current knowledge of molecular nature and the properties of mKATP channels.

Molecular Composition and the Properties of MKATP Channel

Molecular nature of mKATP channel was debated since early 90th after the first mKATP channel discovery by Inoue et al. (1991). Latest discovery of mKATP channel structure in 2019 [51] shed more light on the nature of mKATP channel and confirmed common knowledge on its molecular composition as octameric multiprotein complex. On the whole, molecular architecture of mKATP channel is similar to that of sKATP channel, so we basically can refer to much better studied sKATP channel for brief overview of the structure and the properties of mKATP channel.

Subunit Composition of KATP Channels

Both sKATP and mKATP channels are composed of four pairs of pore-forming K^+ conductant Kir6.x (potassium inward rectifier) subunits and sulfonylurea receptor subunits (SUR). SKATP channels comprise Kir6.1 and Kir6.2 subunits forming the channel pore and SUR1, SUR2A, SUR2B subunits, which accomplish the receptor functions [36, 53]. Experimental evidence was obtained of the presence of Kir6.1, Kir6.2, SUR1, and SUR2A in mitochondria.

SUR subunits are the receptors for physiologicsl ligands (Mg^{2+}, ATP, ADP) and pharmacological modulators of channels activity: potassium channels openers (KCOs) and blockers (primarily sulfonylureas, such as glibenclamide and tolbutamide). Both Kir (Kir6.1 and 6.2) and SUR (SUR1, SUR2A and SUR2B) subunits of KATP channels participate in metabolic regulation of channels activity [53, 54]. The distribution of sKATP and mKATP channels exhibited cell-specificity. Early studies on the distribution of sKATP and mKATP channels have shown the expression of SUR1 in tissues and mitochondria of heart [55], pancreatic beta cells (SUR1 [56]), and liver (where Kir6.1 and SUR1 were found [57]). SUR2A showed relatively high expression in the heart and skeletal muscle [55], whereas SUR2B was more widely expressed in smooth muscle and brain [58].

MgATPase Activity of KATP Channels

SUR subunits of sKATP channels (SUR1 and SUR2, A and B isoforms) belong to the so-called ATP-binding cassette proteins (ABC proteins) containing two nucleotide binding domains forming an ATP-binding pocket [53, 54]. Intrinsic MgATPase activity of SUR, which lacks the transport properties, is important for the regulation of the channel conductance and the activation by KCOs. ATP blocks both sKATP and mKATP channels directly binding to K^+ conductant subunits [53]. SUR plays regulatory role in the mechanism of KATP channels blocking by ATP. In sKATP channels ATP binding site is localized on the cytosolic side of plasma membrane [36, 53]. Convincing evidence was obtained that mKATP channel too was blocked by ATP from the cytosolic side of mitochondrial membrane [59].

Diazoxide Sensitivity of sKATP Channels

Diazoxide (DZ) is most studied and most widely used of mKATP channels openers, and DZ sensitivity is one of the key issues in the understanding of the mechanisms of pharmacological activation of KATP channels. As of sKATP channels, it is known that different pairs of Kir/SUR isoforms differ in their sensitivity towards DZ [56]. While SUR1 comprising channels (e.g., Kir6.1/SUR1 expressed in liver and Kir6.2/SUR1 expressed in beta cells) showed high sensitivity to DZ, SUR2 comprising channels generally were unaffected by this drug [56]. Based on the published data, the difference in the sensitivity to DZ between SUR1 and SUR2 isoforms was critically dependent on Mg·ADP [60], which presence allowed for full activation of otherwise 'insensitive' SUR2A comprising channels at ~300 μM of DZ [60]. Unlike this, Mg·ADP was dispensable for DZ activation of SUR1 and SUR2B comprising channels [53]. From the studies on sKATP channels it was known that KCOs binding to SUR1, SUR2A and SUR2B required MgATPase activity of SUR subunits and a conformational change induced by ATP hydrolysis [53, 60].

Pharmacological properties of mKATP channels were extensively studied by many research groups, and presently are not a subject of much discussion. To understand basic functional consequences of mKATP channels opening we need to outline basic properties of mKATP channels. It worth notion, that all properties of mKATP channel were highly dependent on the preparation used to study mKATP channel activity.

The Blocking of mKATP Channels by ATP

Inoue, et al. [61] first described mKATP channel properties by patch clamping liver mitoplasts. They showed that mKATPchannel was reversibly blocked by ATP (IC_{50} ~ 0.8 mM) from the matrix side. Mg^{2+} was not required for the blocking of the channelby ATP. After half-inhibition by 1 mM ATP, the channel was blocked by 5 mM of 4-aminopyridine (the non-selective blocker of Kv channels) and 5 μM glibenclamide [61]. As showed later Mironova et al. isolated Kir subunit (a 55-kDa protein obtained by ethanol extraction) too was blocked by ATP with low affinity ($K_{1/2}$ ~0.6 mM),and the inhibition did not require the presence of Mg^{2+}, similarlyto mKATP channel studied by Inoue [62]. Meanwhile, in proteoliposomes obtained from liver mitochondria, ATP required the presence of Mg^{2+} for mKATP channel inhibition ($K_{1/2}$ ~22μM), and still much higher affinity for ATP inhibition was observed in intact mitochondria ($K_{1/2}$ ~1μM). So, except isolated Kir subunit and mitoplasts, Mg^{2+} was indispensable for the ATP inhibition of mKATP channel [62]. Also, the requirement for Mg^{2+} and ATP was shown for mKATP channel modulation by pharmacological ligands.

DZ Sensitivity of mKATP Channel

In the works of Garlid et al. high sensitivity of liver and cardiac mKATP channel to DZ (EC_{50} ~3 μM) as opposed to cardiac sKATP channel (EC_{50} ~900 μM), first was found [59, 63]. In early studies this issue was addressed by Liu et al. (2001), who showed different sensitivity to DZ of SUR1 and SUR2 isoforms, and their combinations with Kir subunits of sKATP channels [56]. However, experimentally heart and liver mitochondria did not differ in their sensitivities to DZ [63]. After the recent discovery of mKATP channel [51], this phenomenon can be explained by different nature of the channel; however molecular basis of mKATP channel activation by DZ still is unclear.

DZ activation showed strong dependence on mKATP channels preparations. Thus, $K_{1/2}$ for DZ activation of reconstituted mKATP channel was as low as ~0.37 μM, while DZ affinity of mKATP channel in mitochondria was ten times weaker ($K_{1/2}$ ~3-4 μM) [63]. As it was stressed in Garlid's works, independent on the preparations, there was an absolute requirement for Mg^{2+} and ATP for the opening of mKATP channelby DZ.

MKATP Channel Blocking by Glibenclamide

The sensitivity of mKATP channel to the blocking by glibenclamide too was highly dependent on the preparations of mKATP channels. In proteoliposomes [59], glibenclamide was capable of blocking mKATP channel in the absence of ATP with high affinity ($K_{1/2}$ ~0.25 μM). The presence of ATP increased the affinity of reconstituted channel to glibenclamide ($K_{1/2}$~0.09 μM). In mitochondrial preparations, thechannel affinity to glibenclamide blocking was by the order lowerthan in proteoliposomes ($K_{1/2}$ ~1-6 μM). However, several research groups stressed that neither glibenclamide, nor selective mKATP channel blocker 5-HD werecapable of blocking mKATP channel in the intact mitochondriainthe absence of Mg^{2+} and ATP [2, 59, 62, 63]. So, the blockage of mKATP channel first required its opening by DZ (or other KCOs) in the presence of Mg^{2+} and ATP, which enabled the blockage of activated channel by glibenclamide or 5-HD [59, 62]. This was explained by the conformational changes making pharmacologically openedmKATP channel susceptible to the blocking by pharmacological agents [59, 63].

From all the above data it followed that membrane environment was of importance in the regulation of mKATP channel affinity to physiological and pharmacological ligands. However, till present the proteins involved in the regulation of mKATP channels activity were not identified. One questionable issue in the knowledge of mKATP channelsproperties was the requirement for Mg^{2+} and ATP for mKATP channel interactions with pharmacological modulators. Since the early works of Garlid's group, it was stated that the presence of MgATP complex was obligatory for any effect of KCOs (diazoxide, pinacidil), or the blockers (glibenclamide, 5-HD) on mitochondrial K^+ transport and mitochondrial functions: respiration, matrix swelling, Ca^{2+} uptake, and ROS production [2]. However, literary data didn't show consensus regarding this issue.

Ability of KCOs (diazoxide, pinacidil, cromakalim) to increase, and of mKATP blockers (glibenclamide, 5-HD) to blockpotassium transport in the intact isolated mitochondria without MgATP was shown in several works including our research. Susceptibility of mitochondrial ATP-sensitive K^+

transportto the activation by KCOs without MgATP was observed in different tissues: heart [43], skeletal muscle [64], and liver [65]. Similarly,in the absence of MgATP glibenclamide and 5-HD blocked K^+ transport related to mKATP channel activity in skeletal muscle [64], liver [65], and brain[66]. In our works, glibenclamidetoo was effective as the blocker of ATP-sensitive K^+ transport without MgATP at concentrations assumed to be 'safe'for mitochondria (~5-10 μM). The above phenomena weretoo ubiquitous to consider it as an artifact, coincidence, or merely off-target effects. However, it is still unclear, whether described phenomena reflect the effects of the drugs on mKATP channel activity, and if so, which molecular entity in mitochondrial membrane represents ATP-sensitive K^+ transport susceptible to the activation by KCOs andthe blockage by mKATP channels blockers in the absence of MgATP.

MKATP Channel: The Evidence and the Hypotheses

Recent discovery of mKATP channel in 2019 [51] based on combined proteomic, genetic, biophysical and biochemical approaches revealed molecular nature of mKATP channel, but put still more questions about its physiological relevance. On the whole, an octameric structure of mKATP channel comprising four pairs of K^+ conductant and receptor subunits named mitoK and mitoSUR was confirmed. The channel exhibited biophysical and biochemical properties known of mKATP channels. Till present there was little evidence yet on the physiological functions of mitoK, but it was shown that mitoK was indispensable for the maintenance of mitochondrial $\Delta\Psi_m$ and the ability of mitochondria to ATP synthesis [51].

The novel discovery of mKATP channel gave the researchers a powerful tool to reveal molecular basis underlying the role of mKATP channels in tissue protection under several diseases and pathophysiological conditions [4]. This started a new wave in mKATP channels research on molecular level. However, now it needs to be answered how many KATP channel types are present in mitochondria, and which physiological functions are accomplished by these channels. Recent studies confirmed the requirement of mitochondrial Kir6.1 and Kir6.2-KATP channels for cardio- and neuroprotection [11, 14, 30]. So, which are physiological functions of mitoK and Kir6.1, Kir6.2-KATP channels expressed in mitochondria? To answer this question, we hypothesize that these channels, having similar K^+ conductant properties, could take part in different biological processes, not yet fully identified.

In recent works, the novel role of mKATP channels in the regulation of mitochondrial functions and dynamics was found. So, it was shown that both mitoK- and Kir6.x- KATP channels were indispensable for cardio- and neuroprotection. Particularly, in brain the antidepressive effects of mKATP channels opener iptakalim was shown, and the involvement of K^+ conductant subunit (mitoK) in the improvement of synaptic plasticity in mice was found [12]. Kir6.1/K-ATP channel functioning in brain too was protective against neurodegeneration in the MPTP model of Parkinson's disease by promoting mitophagy [14]. Kir6.2 knockout promoted morphological impairments in brain mitochondria, inhibition of ATP synthesis, and behavior impairments in mice [11]. In the heart, recently it was shown that Kir6.1 knockout aggravated cardiac dysfunction in diabetic cardiomyopathy, and Kir6.1 overexpression improved cardiac function, which confirmed essential role of Kir6.1 in cardioprotection [67]. Earlier research [68] has shown that Kir6.2 was required for cardioprotection by IPC. Kir6.2/SUR2A channels protected cardiomyocytes against anoxia/reoxygenation *via* Akt/GSK-3β and Akt/mTOR signaling pathways [30]. An interaction of mitochondrial Kir6.2 with Hsp90, and the interaction of Kir6.1 with connexin 43 were required for the protection of cardiomyocytes against hypoxic injury [46, 47]. These facts are too numerous to rule out physiological relevance of Kir6.x-KATP channels in mitochondria.

It is reasonable to suppose that besides K^+ conductant properties, mKATP channels of mitochondrial membrane could specifically interact with not yet identified membrane proteins in order to maintain vital biological functions. Novel evidence of vital role of mitoK, kir6.1/KATP and kir6.2/KATP channels in the maintenance of mitochondrial functions confirmed the hypothesis of the multiplicity of mKATP channels and indicated variety of their cellular functions.

Bioenergetics of K^+ transport first of all requires the knowledge of the abundance of the K^+ channels in mitochondrial membrane, the estimates of the rate of ATP-sensitive potassium transport, its share in oxygen consumption and K^+ cycle, and its impact on mitochondria energy state. It is worth noting that all mKATP channels types were sensitive to mKATP channels openers: diazoxide, nicorandil, iptakalim [12, 13, 51]. Also, these channels exhibited similar biophysical properties, and their opening in mitochondria had similar bioenergetic consequences. This partially allows for abstracting of cellular and molecular mechanisms of mKATP channels activation and functioning. The aim of this work was an overview of our data on the basic properties of ATP-sensitive K^+ transport and its bioenergetics effects in isolated mitochondria.

The Properties of ATP-Sensitive K^+ Transport in Mitochondria

Despite extensive studies of the properties of mKATP channel, in the literature there was controversial data regarding the role of Mg^{2+} and ATP in the mechanisms of the regulation of channel activity by pharmacological modulators. In our works we have shown that ATP-sensitive K^+ transport in liver and brain mitochondria was activated by diazoxide and pinacidil on sub-micromolar scale in the absence of MgATP. This indicated that MgATPase could be dispensable for mKATP channel activation by KCOs [69]. The effects of DZ and pinacidil on K^+ transport were assessed indirectly based on the oxygen consumption and light scattering of mitochondrial suspensions. As it is known, oxygen consumption is coupled to K^+ transport by stoichiometric proportions, and under proper conditions increase in the rate of respiration is proportional to the rate of K^+ uptake. For NADH-dependent substrates K^+: O ratio constitutes 10:1 [70], which allows for quantitative estimation of K^+ transport. Also, because of obligatory water uptake, the rate of matrix swelling too reflects the rate of K^+ uptake [71].

In the absence of MgATP DZ and pinacidil increased both the rate of oxygen consumption and the amplitude of matrix swelling in brain and liver mitochondria which reflected the elevation of K^+ uptake (Figure 1A, B). This occurred in a concentration-dependent way with maximum at <0.5 μM (Figure 1C, D). The rate of oxygen consumption, as well as swelling of brain mitochondria did not increase with the increase of KCOs concentration above 0.5 μM (Figure 1C, D). As for liver mitochondria, recently we have shown that DZ was capable of non-specific activation of K^+ transport, which was abolished in the presence of Mg^{2+} [69]. This effect was not observed in brain mitochondria, which made the preparations of brain mitochondria more suitable to study the sensitivity of mKATP channel to KCOs.

The Sensitivity of Mitochondrial K^+ Transport to KCOs

As showed normalized concentration dependences of the rate of oxygen consumption and the initial rates of matrix swelling (V_0), in the absence of MgATP half activation of K^+ transport of both liver and brain mitochondria occurred on sub-micromolar scale. EC_{50} for DZ activation of K^+ transport constituted ~150 nM and ~180 nM in liver and brain mitochondria (Figure 1C, D). Either Mg^{2+} or ATP alone at concentration 1 mM had no effect on the activation of K^+ transport in native mitochondria by DZ and pinacidil (Figure 1F). ATP did not block K^+ transport in the absence of Mg^{2+}, which agreed with the literature [62].

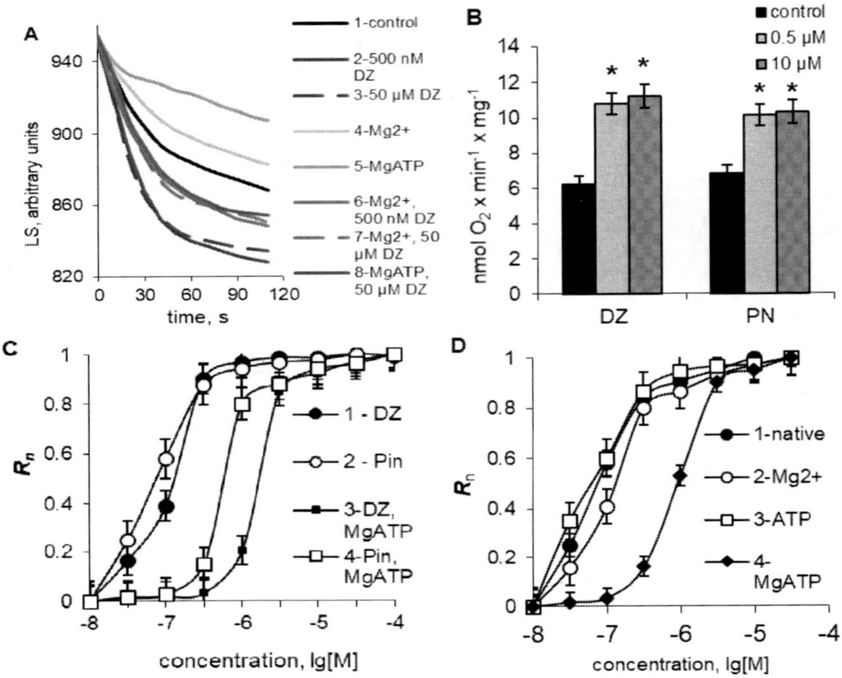

Figure 1. The effects of diazoxide and pinacidil on light scattering (A) and oxygen consumption (B) of brain mitochondria. C, D: the sensitivity of K+ transport to DZ and pinacidil, concentration dependences in the absence and the presence of MgATP [72]. Incubation medium: 150 mM KCl, 5 mM glutamate, 1 mM Pi, 0.5 mM EDTA. (EGTA), Mg^{2+} was added at 1 mM, ATP was added at 0.3 mM, as indicated. M ± m, n = 6; * - $P < 0.05$ as compared to control.

In the presence of MgATP, conventional shift of EC_{50} towards much higher concentrations of KCOs was observed, with full activation at ~1-3 µM (Figure 2C, E). In native mitochondria the activation of K^+ transport by nanomolar concentrations of DZ and pinacidil was not observed in the presence of either MgATP (Figure 1D), or KATP channels blockers, glibenclamide and 5-HD (*not shown*), which indicated the sensitivity of the activation effect to the blockers of mKATP channel. In the presence of MgATP, glibenclamide and 5-HD conventionally blocked K^+ transport activated by 30 µm DZ (Figure 1D) and 10 µM pinacidil (Figure 2A, B).

So, without MgATP, we obtained strong evidence of high sensitivity of ATP-sensitive K^+ transport to diazoxide and pinacidil with full activation at <0.5 µM of the drugs. MgATP shifted it to micromolar concentration level, which agreed with literary data. By our earlier estimations, in native brain and liver mitochondria V_0 of ATP-sensitive K^+ transport constituted ~40 nmol $K^+ \cdot min^{-1} \cdot mg^{-1}$. In liver mitochondria V_0 of ATP-sensitive K^+ transport rose to ~65 nmol $K^+ \cdot min^{-1} \cdot mg^{-1}$ upon activation by DZ [69]. In brain mitochondria DZ increased V_0 of ATP-sensitive K+ transport by ~1.5-2 times, so that upon activation by DZ it could reach up to ~80 nmol $K^+ \cdot min^{-1} \cdot mg^{-1}$.

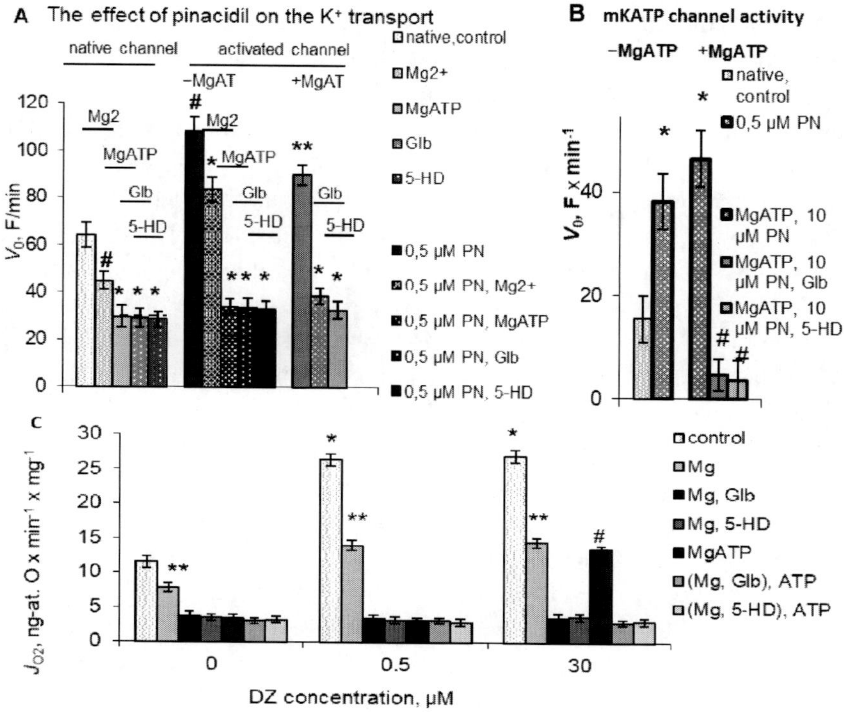

Figure 2. The effects of DZ and pinacidil on ATP-sensitive K^+ transport in the presence and the absence of MgATP [72]. The additions are shown on the legends. M ± m, n = 6; * - P < 0.05 as compared to control.

Pharmacological Identification of ATP-Sensitive K^+ Transport

One crucial point of our study was the attribution of ATP-sensitive K^+ transport to mKATP channels activity. Generally, identification of mKATP channel, especially after its novel discovery [51], required genetic approaches. However, in our study we made an attempt to identify ATP-sensitive K^+ transport on pharmacological level using DZ, pinacidil, and the blockers of KATP channels, glibenclamide and 5-HD.

For this purpose we used polarographic method which allowed for sequential blocking of ATP-sensitive K^+ transport by different blockers under steady-state conditions (state 4 respiration). Using this approach we have shown the identity of K^+ transport, blocked by glibenclamide and activated by DZ in the absence of MgATP, with K^+ transport conventionally activated and blocked in the presence of MgATP.

Experiments were conducted in the presence of Mg^{2+}, because this allowed for the direct comparison between the parts of K^+ transport blocked by glibenclamide (5-HD) and MgATP. Also, consistent with literature [59], we assumed that MgATP was capable of blocking mKATP channel in native mitochondria. As we have found, part of K^+ transport blocked by glibenclamide equaled to K^+ transport blocked by ATP; after the blocking by glibenclamide no additional blocking by MgATP was observed. The same was observed with 5-HD (Figure 2). This proved the identity of K^+ transport blocked by glibenclamide and MgATP and the ability of glibenclamide to block K^+ transport without MgATP. Also, in the presence of Mg^{2+}, but the absence of ATP, K^+ transport activated by KCOs was similarly blocked by glibenclamide and MgATP. K^+ transport blocked by glibenclamide and 5-HD equaled to K^+ transport blocked by MgATP and to the K^+ transport conventionally activated by DZ and pinacidil in the presence of MgATP (Figure 2). This proved the identity of K^+ transport sensitive to DZ, pinacidil, glibenclamide, and MgATP, and justified its attribution to mKATP channels activity. While in sKATP channels MgATPase activity was required for the channel opening by DZ and pinacidil, our data on liver and brain mitochondria showed that MgATPase activity was dispensable for mKATP channels activation by these drugs. MgATPase activity as well was dispensable for mKATP channel blocking by glibenclamide and 5-HD, however, the channel blocking by ATP required the presence of Mg^{2+} [72].

Based on the experiments we came to the following conclusions: 1) high sensitivity to DZ and pinacidil with full activation at $<0.5\ \mu M$ independent of MgATPase activity was one common property of native mKATP channels in isolated mitochondria; 2) Mg^{2+} alone could not affect the mKATP channels affinity to KCOs, but the presence of MgATP shifted it to much higher micromolar concentrations; 3) the results of our work allowed us assume that native mKATP channel might comprise the sites with high affinity to diazoxide and pinacidil, possibly screened by the binding of MgATP. Obtained results indicated novel common features in the mechanism of native mKATP channel activation by pharmacological KCOs.

Bioenergetics of ATP-sensitive K^+ Transport in Mitochondria

The Contribution of mKATP Channel to the K^+ cycle

The assessment of bioenergetics consequences of mKATP channel opening (its impact on mitochondrial energy state, Ca^{2+} transport, ATP synthesis, and ROS production) needs to prove its participation in the K^+ cycle and to find the share of ATP-sensitive K^+ transport in oxygen consumption. Generally, potential-dependent K^+ uptake in the presence of permeable anion (phosphate) is accompanied by obligatory water uptake and matrix swelling. Lowering of Mg^{2+} concentration in the matrix and dissociation of Mg^{2+} ions from Mg^{2+} binding sites demasks the activity of of K^+/H^+-exchanger and eventually results in the activation of K^+ cycle: potential-dependent K^+ uptake and K^+ efflux via K^+/H^+-exchanger [71]. K^+ cycle is an energy dissipating process and potential-dependent K^+ uptake leads to the obligatory increase in oxygen consumption.

To assess the share of mKATP channel in the K^+ cycle, we developed an approach based on the recording of the initial rates (V_0) of K^+ transport after the K^+ uptake and the establishment of an equilibrium between K^+ uptake *via* K^+ channels and the efflux *via* K^+/H^+-exchanger, which altogether constitute K^+ cycle and corresponds to oxygen consumption in state 4 [71]. The functioning of K^+ cycle allowed for keeping the matrix volume at equilibrium, dependent on the equilibrium between the K^+ uptake and the efflux. This was confirmed by the blockage of the K^+ uptake, which resulted in the equilibrium shift towards the K^+ efflux and matrix contraction.

Based on this reasoning, mKATP channels activity corresponded to the difference between the initial rates of matrix contraction before and after selective blockage of mKATP channel by glibenclamide: $V_{KATP} = V_1 - V_0$ (Figure 3, A, 1-2). The share of potential-dependent K^+ channels in K^+ cycle was estimated similarly as the difference between V_0 of matrix contraction after the non-selective blocking of K^+ channels with 4-AP and the blockage of mKATP channel by glibenclamide: $V_K = V_2 - V_1$ (Figure 3A, 2, 3). For the activation of mKATP channel we used DZ as best studied of mKATP channels openers. The activation of K^+ transport by DZ resulted in the activation of K^+ cycle (Figure 3A, B) and the increase in the rate of respiration (Figure 3C, D). Our approach allowed for the evaluation of the share of native and activated mKATP channel in the K^+ cycle in the absence and the presence of DZ (Figure 3B, D). The activation of the K^+ cycle was confirmed by monitoring proton and K^+ transport with the probes BCECF and PBFI [69].

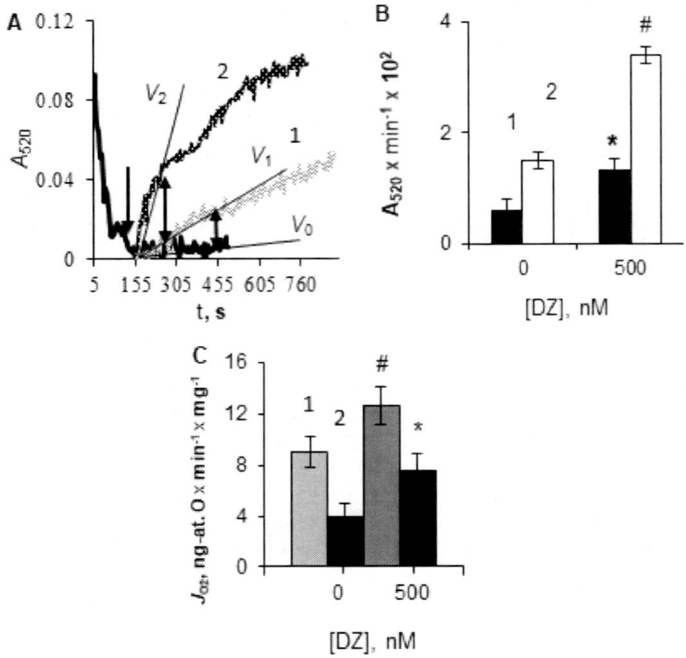

Figure 3. The activation of potassium cycle by DZ {the details are in the text). A: absorbance after rthe additions of glibenclamide (1), glibenclamide+4-AP (2); B: glibenclamide-sensitive (1) and total (2) amplitude of matrix shortage; C: ATP-insensitive (1) and ATP-sensitive (2) components of oxygen consumption. M ± m, n = 6; *, # - $P < 0.05$ as compared to control.

As it was shown in our works, all bioenergetics effects produced by DZ as the activator of K^+ transport occurred on sub-micromolar scale and corresponded to the activation of mKATP channel by this drug with EC_{50} ~140 nM in liver mitochondria [65, 69, 72]. As we have found in liver mitochondria, DZ was capable of activation of not only mKATP channel, but ATP-insensitive component of K^+ transport (Figure 3B, D), which was completely blocked by Mg^{2+} [69]. However, for the purpose to assess bioenergetic effects of mKATP channel opening by DZ, we conducted experiments in the absence of MgATP because of well-known inhibition of K^+/H^+-exchanger and K^+ cycle by Mg^{2+} ions. So, our approach to evaluation of bioenergetics effects of ATP-sensitive K^+ transport was based on the finding of the share of mKATP channel in oxygen consumption and mitochondrial uncoupling stimulated by DZ (Figure 4).

Bioenergetic Effects of mKATP Channel Opening by DZ: Mild Uncoupling of Ca^{2+} Transport, ROS Production, and Phosphorylation

As we observed in our studies, bioenergetic effects of DZ in liver mitochondria largely could be explained by 'classical' mild uncoupling, which occurred without membrane depolarization. From the literature, it was known that $\Delta\mu_H$ (and consequently $\Delta\Psi_m$) falls linearly with the increase in the rate of state 4 respiration [73]. This allows for the evaluation of depolarization effect by measuring the impact of ATP-sensitive K^+ transport on the rate of respiration. In brain mitochondria, the estimation of the share of mKATP channel in the respiration gave a depolarization effect of about ~20% of $\Delta\Psi_m$, proportionally to the contribution of ATP-sensitive K^+ transport to state 4 respiration [66]. In heart mitochondria, the estimateof ~1-2 mV of depolarization [73] was within experimental error when $\Delta\Psi_m$ was assessed using conventional techniques (potentiometry with TPP+ electrode or fluorescent probes). As we observed in liver mitochondria, the rate of K^+-stimulated respiration, even at full activation of K^+ transport by DZ, too was insufficient to cause significant depolarization effect [65]. For this reason, liver mitochondria were more suitable, as compared to brain mitochondria, for the study of the effects of 'mild uncoupling' on energy-dependent processes in mitochondria.

Ca^{2+} Transport

In the literature it was often assumed that depolarization was the main cause of the reduction of Ca^{2+} uptake by KCOs, and very high doses of DZ were required for non-specific depolarization of heart mitochondria, not caused by mKATP channel opening, and the inhibition of Ca^{2+} transport [73-75]. However, in our work on liver mitochondria, the reduction of Ca^{2+} uptake by as low as <0.5 μM of DZ could be explained by the 'classical uncoupling' without apparent changes in $\Delta\Psi_m$. The effect of K^+ transport on Ca^{2+} uptake was studied by the additions of the same aliquots of Ca^{2+} with the increasing K^+ concentration in the medium in the absence and the presence of DZ. V_0 of Ca^{2+} uptake decreased with the increase of K^+ concentration and the rate of K^+ uptake (Figure 4A). As we observed earlier, DZ reduced both V_0 of Ca^{2+} uptake and Ca^{2+} accumulation capacity of liver mitochondria [76].

In liver mitochondria, there was an inverse correlation between K^+ and Ca^{2+} uptake (Figure 4A). So, we supposed that the reduction of Ca^{2+} uptake was caused by the overall ability of mitochondria for cation uptake, which was limited by the rate of respiration. So, uncoupling effect of K^+ appeared as the reduction of Ca^{2+} uptake at simultaneous elevation of K^+ uptake, so that a total rate of positive charge transfer (q^+) remained relatively stable (Figure 4B).

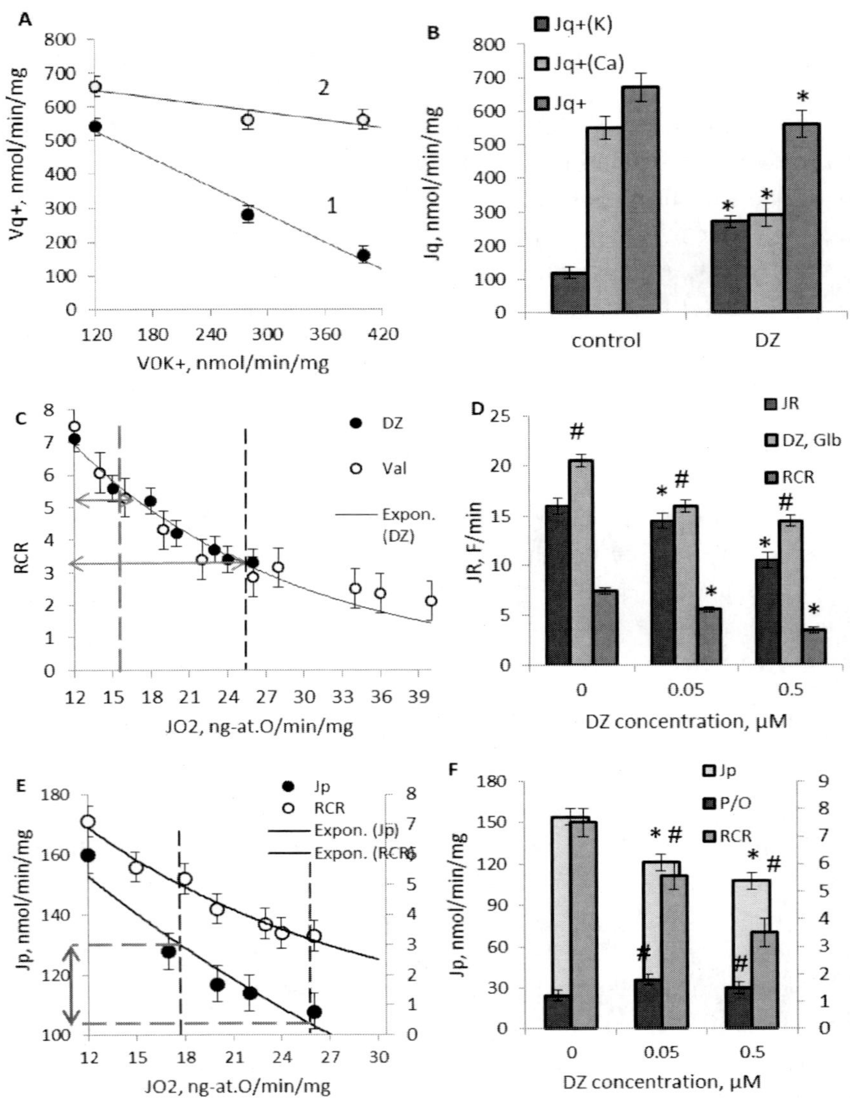

Figure 4. The effects of mild uncoupling of the respiratory chain by diazoxide on Ca^{2+} transport (A, B), ROS production (C, D), and ATP synthesis (E, F). Details are in the text. M ± m, n = 6; *, # - $P < 0.05$ as compared to control.

This was confirmed by our estimates of the rates of cation transport: K^+ uptake, Ca^{2+} uptake and the total rate of positive charge transfer (Figure 4B). So, under the conditions of full activation of K^+ transport by DZ, the rate of total charge transfer constituted 560 nmol $q^+ \cdot min^{-1} \cdot mg^{-1}$, which was only ~1.2 times lower than control 670 nmol $q^+ \cdot min^{-1} \cdot mg^{-1}$, in spite of almost twofold reduction of V_0 of Ca^{2+} uptake (1.9 times). While we need to analyse the effects of DZ on Ca^{2+} transport in more details, the above estimate could explain the limiting effect of K^+ on Ca^{2+} uptake, which unlike concurrent inhibition, showed weakening of inhibitory effect with the lowering of Ca^{2+} concentration, and *vice versa*, stronger inhibition with the increase of Ca^{2+} concentration [76].

The Effect of Mild Uncoupling on ROS Production

In liver mitochondria respiring on glutamate, the activation of K^+ cycle eventually resulted in the decrease of RCR in a measure of the increase of the rate of diazoxide-stimulated respiration. By using valinomycin, we confirmed that uncoupling was caused by K^+ transport, because similar stimulation of respiration resulted in similar reduction of RCR (Figure 4C). Thus, the absence of depolarization effect in liver mitochondria made DZ a very useful tool for the study of bioenergetic effects of K^+ transport produced 'purely' by the mild uncoupling of the respiratory chain. Knowing the share of mKATP channel in diazoxide-stimulated respiration (Figure 4C), we estimated the share of fully activated mKATP channel in mitochondrial uncoupling, which resulted in the decrease of RCR by ~2 units (Figure 4C, dashed lines). Based on the correlation dependence between the decrease in RCR caused by the stimulation of K^+ cycle and the reduction of ROS formation caused by forward electron transport (*not shown*), it was possible to evaluate the share of mKATP channel in the reduction of ROS formation by DZ (Figure 4D). Accordingly, the dependence of both RCR and the rate of ROS formation on the rate of diazoxide-stimulated respiration could fully explain the effect of glibenclamide, which restored ROS production by blocking the part of diazoxide-stimulated respiration related to mKATP channel activity (Figure 4D). Similarly, we have shown that glibenclamide restored $\Delta\Psi_m$ and ROS production in brain mitochondria proportionally to the blockage of the share of mKATP channel in state 4 respiration [66].

Oxidative Phosphorylation

Uncoupling of the respiratory chain by diazoxide affected Oxphos system as well. Stimulation of ATP-sensitive K^+ transport by DZ slowed down the rate of phosphorylation, however, up-regulated its effectiveness, P/O (Figure 4E, F). Our data obtained polarographically well agreed with our earlier results obtained by monitoring proton transport [20]. By our estimation (Figure 4E), the functioning of fully activated mKATP channel reduced the rate of phosphorylation by ~30 nmol ADP·min^{-1}·mg^{-1}. ATP hydrolysis too was inhibited by DZ, which indicated more complex mechanism of DZ action on F_0F_1 ATP synthase than merely mitochondrial uncoupling.

As showed published data, including the results of our studies, state 3 respiration and ATP synthesis were directly suppressed by mKATP channels openers, diazoxide [20, 43, 64], cromakalim [74], and pinacidil [64], however, this phenomenon was not fully explained in the literature. Non-specific inhibition of F_0F_1 ATP synthase should be ruled out because it required much higher concentrations of DZ, and similar effects were produced by chemically different drugs. So far, we propose some considerations aimed to explain observed contradictory effects – the inhibition of oxidative phosphorylation, and the improvement of phosphorylation efficiency caused by the activation of mKATP channel.

Thus, we suppose that activation of ATP-sensitive K^+ transport by DZ, besides uncoupling, could inhibit phosphorylation by the activation of K^+/H^+ exchange and K^+ cycle. Based on literary data, there was a possibility that activation of K^+ and H^+ fluxes could intervene with molecular mechanism of F_0F_1 ATP synthase functioning. As it is known presently, enzymatic activity of F_0F_1 ATP synthase involves both proton and K^+ translocation across mitochondrial membrane [52]. It is reasonable to suppose that K^+ uptake and the activation of K^+/H^+-exchange could disturb the coupling between K^+ and H^+ transport, and ATP synthesis. Possibly, similar uncoupling of the functioning of F_0F_1 ATP synthase at molecular level by the activation of K^+/H^+-exchange was described in much earlier study of the effects of K^+/H^+-ionophore gramicidin on F_0F_1 ATP synthase activity [77]. Such mechanism called decoupling of F_0F_1 ATP synthase occurred without apparent changes in $\Delta\mu_H$, however, strongly suppressed ATP synthesis [77]. While this mechanism requires much more detailed study, it is tempting to hypothesize that a sort of decoupling of F_0F_1 ATP synthase by the activation of K^+/H^+ exchange caused by mKATP channels opening could contribute to the inhibition of ATP synthesis and hydrolysis observed experimentally.

Other remarkable feature of the inhibition of ATP synthesis by mKATP channels opening was simultaneous down-regulation of ATP production and the increase of phosphorylation efficiency (P/O). At present we have not enough data yet to explain an increase in P/O caused by mKATP channels opening. However, it is worth mentioning that in the literature, "mild" respiratory uncoupling was shown to increase P/O ratio [78]. Thus, tightly coupled mitochondria exhibited low P/O ratio because of higher proton leak as compared to partially uncoupled ones [78]. This could explain lower P/O found in our work in control mitochondria, which exhibited tight coupling and high RCR in the absence of mKATP channel opener (Figure 4F). So, mild uncoupling by diazoxide, possibly, created conditions optimal to reach higher phosphorylation efficiency at lower RCR (Figure 4F).

From Bioenergetics to Functions: The Effect of Diazoxide on MPTP Opening in Liver Mitochondria

Mitochondrial permeability transition pore (mPTP) is one acknowledged principal target of cardioprotective therapy using pharmacological mKATP channels openers [1, 2]. MPTP is a non-selective mitochondrial megachannel built up of proteins of inner and outer membrane, which opening is triggered by ROS and calcium overload of mitochondria (reviewed in [5, 79]). The inhibition of mPTP activity following mKATP channels opening was shown to prevent the development of cellular apoptosis and necrosis under pathophysiological conditions [1, 7]. The opening of mKATP channels can suppress mPTP activity by several mechanisms. At cellular level, mPTP inhibition by mKATP channels openers occurs both indirectly, by triggering signaling pathways [1, 4], and more directly by the modulation of ROS production and Ca^{2+} uptake in mitochondria. However, because of a great diversity in the published data about the direct effects of mKATP channels opening on ROS production (both activation [1, 31, 38], and inhibition [2, 15-17] were shown to be cytoprotective in cardiovascular and central nervous systems), no consensus was reached regarding the mechanisms regulating mPTP activity by mKATP channels opening.

As we have shown earlier, even full mKATP channel activation by DZ did not prevent accumulation of low amounts of Ca^{2+} [76]. So, it was of interest to study the effects of DZ on mPTP activity under the conditions when Ca^{2+} uptake was unaffected by ATP-sensitive K^+ transport. Ca^{2+} uptake was controlled spectrophotometrically with Ca^{2+} indicator arsenaso III. MPTP opening was triggered by the additions of as low amounts of Ca^{2+} as 10 μM. Under these conditions mPTP could operate in one of its sub-conductance states, and take part in Ca^{2+} cycling without depolarization of mitochondrial membrane [80]. MPTP activity was assessed spectrophotometrically following the changes in equilibrium matrix volume (V_m). As we have shown, the inhibition of ROS production caused by the stimulation of K^+ cycle was a sufficient requirement for the suppression of mPTP activity in liver mitochondria.

The Share of mPTP in the Regulation of Matrix Volume Upon the Activation of mKATP Channel

To evaluate the contribution of mPTP and mKATP channels in the regulation of matrix volume, we applied the same approach as we used in the study of the K^+ cycle [80]. Under the conditions of concurrent functioning of mPTP and mKATP channels an equilibrium matrix volume was regulated by the functioning of mPTP, K^+ channels, and K^+/H^+-exchanger. The purpose of our work was to assess the share of mPTP in the regulation of matrix volume, and evaluate how it was affected by DZ under steady-state conditions reached after the accumulation of Ca^{2+} and K^+ into matrix, and the equilibration of K^+, Ca^{2+}, and H^+ fluxes across mitochondrial membrane.

The functioning of mPTP and K^+ channels led to the increase of matrix volume, whereas the work of K^+/H^+-exchanger led to matrix contraction. The blocking of K^+ channels (by glibenclamide and 4-AP) increased the rate of matrix shortage by the value equal to the contribution of K^+ channels to the equilibrium matrix volume. Based on this reasoning, the share of mPTP in the regulation of matrix volume (V_m) was found as the difference of initial rates of matrix shortage after the blocking of K^+ channels in the absence and the presence of mPTP blocker cyclosporine A: $V_{MPTP} = V_2 - V_1$ (Figure 5A). The share of K^+ channels in the regulation of V_m was found as the difference between V_0 of matrix contraction before and after the addition of K^+ channels blockers: $V_K = V_1 - V_0$ (Figure 5A).

As we have found, the activation of mKATP channel led to mPTP inhibition and the reduction of its share in the regulation of V_m. This was observed as the reduction of cyclosporine-sensitive and the elevation of cyclosporine-insensitive components of matrix contraction after the blocking of K^+ channels (Figure 4A, B). As we have shown earlier [80], inhibitory effect increased with the increase in DZ concentration and the activation of K^+ cycle (Figure 5C). MPTP opening largely contributed to the activation of K^+ cycle. Elimination of Ca^{2+} transport by EDTA after mPTP opening demasked full scale of the activation of K^+/H^+-exchange (Figure 5A, B), which rate increased from 23.0 ± 4 nmol $H^+ \cdot min^{-1} \cdot mg^{-1}$ in the absence of Ca^{2+} in the medium to 60.0 ± 7 nmol $H^+ \cdot min^{-1} \cdot mg^{-1}$ after mPTP opening, as we have found with the probe BCECF [80].

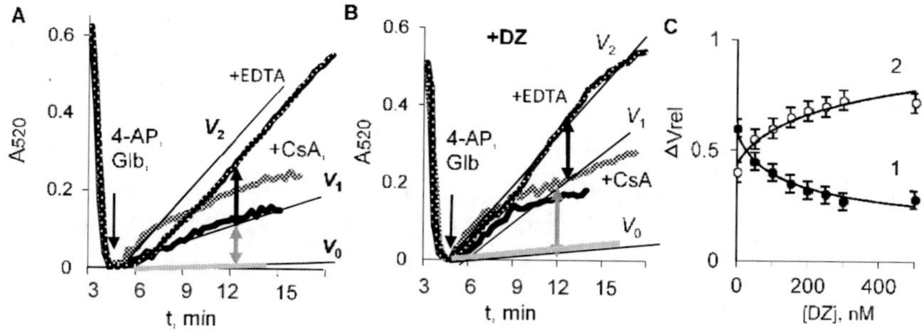

Figure 5. The effect of DZ on matrix volume in mitochondria under state 4 respiration. The additions of cyclosporine A and EDTA are indicated; A, B: the additions of K^+ channel blockers are shown by the arrows (details are in the text). C: the effect of DZ on relative contributions of cyclosporine-sensitive and cyclosporine-insensitive components to the total rate of matrix shortage. M ± m, n = 6.

To explain mPTP inhibition by mKATP channels opening under the conditions of our experiments, it worth mention that mPTP is highly sensitive to ROS production [2, 5, 79]. So, under the conditions of similar Ca^{2+} uptake, the reduction of ROS formation by mKATP channel activation could be most relevant explanation for the inhibition of mPTP activity by DZ (Figure 3A). It was unlikely that mPTP suppression by mKATP channels opening could be mediated by the activation of mitochondrial PKCε [1, 38], because PKCε activation required alcalinization of mitochondrial matrix and the increase in hydroperoxide formation [38]. In our works on liver mitochondria, on the contrary, considerable decrease in ROS production, and fast matrix acidification due to the activation of K^+/H^+ exchange disagree with an assumption of possible PKCε involvement in mPTP inhibition. So, bioenergetic effects of mKATP channels opening: the respiratory uncoupling and the suppression of ROS production were sufficient requirements for mPTP inhibition by DZ in isolated liver mitochondria.

Functional Consequences of mKATP Channels Opening *in Vivo*: The Impact of mKATP Channels Opening on the Physical Endurance

As compared to mPTP channel, the system of oxidative phosphorylation, received much less attention as the target of the pharmacological modulators of mKATP channels. As we have shown, Oxphos is a direct target of ATP-sensitive K^+ transport *in vitro*. However, for biochemistry and physiology it is important to assess physiological relevance of bioenergetics effects of mKATP channel opening. In our works we have shown that modulation of the Oxphos system by ATP-sensitive K^+ transport directly correlated with physical endurance of the animals subject to compulsory swimming with a load. It is a common notion that ATP production is a key issue for the bioenergetics of the living organism. Intensive ATP consumption under physical exercise is accompanied by ATP synthesis and massive oxygen consumption by mitochondria, which causes a sort of oxygen shortage. So, 'optimization' of ATP synthesis and oxygen consumption is critical for the maintenance of a living organism under hypoxic conditions caused by physical loads.

Earlier it was established that the animals differed in their resistance to acute hypobaric hypoxia and physical stress caused by swimming with a load until fatigue [81, 82]. According to published data, different resistance to physical loads also was accompanied by the great alterations in metabolism, primarily at mitochondria level. Recently, it was shown that animals differing in their endurance to exercise training, exhibited higher content of respiratory chain complexes [82], higher respiratory reserve, higher resistance to mPTP opening, higher lipids-over-glucose utilization, and by ~30% prolonged lifespan [83].

By our earlier observations [20], increase of mKATP channel activity caused by short-term acute hypobaric hypoxia resulted in the improvement of the rats' endurance to physical stress. So, it was of interest to examine possible relationship between physical endurance of the rats subject to compulsory swimming with a load and endogenous activity of mKATP channel.

For this purpose, male Wistar rats were subject to compulsory swimming till fatigue and, based on their swimming time (ST), were separated in two groups with high and low resistance to physical stress: HR and LR groups. MKATP channel activity was determined *ex vivo* with K^+-selective microelectrode [84]. Mitochondrial functions were assessed polarographically. Using K^+-selective electrode, we have found that the rate of ATP-sensitive K^+ transport related to mKATP channel activity constituted 40.0 ± 4.0 nmol $K^+ \cdot min^{-1} \cdot mg^{-1}$ in HR group, while LR animals exhibited only 28.0 ± 3.0 nmol $K^+ \cdot min^{-1} \cdot mg^{-1}$.

So, endogenous mKATP channel activity strongly correlated with the resistance of the rats to physical stress in such a way that HR animals exhibited elevated activity of mKATP channel as compared to LR ones (Figure 1A,B). To examine further the relationship between mKATP channel activity and physical endurance, animals of HR and LR groups were separated into three subgroups treated respectively with glibenclamide, 5-HD, and with no treatment but physiological solution, and subject to compulsory swimming. MKATP channel activity was estimated *ex vivo* after the experiment.

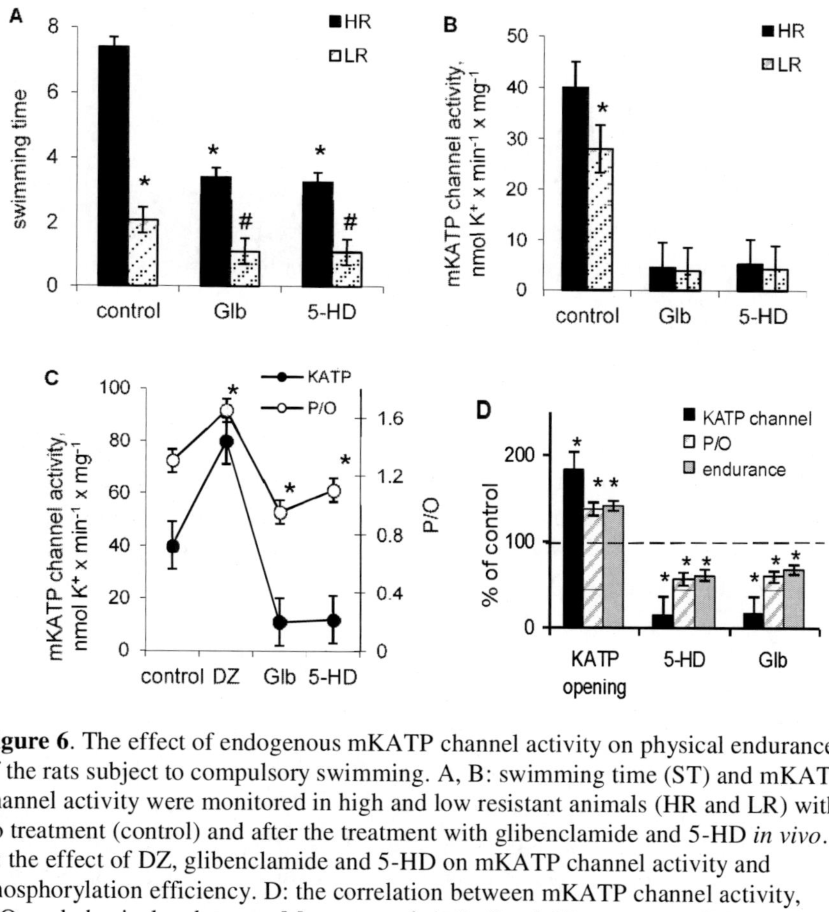

Figure 6. The effect of endogenous mKATP channel activity on physical endurance of the rats subject to compulsory swimming. A, B: swimming time (ST) and mKATP channel activity were monitored in high and low resistant animals (HR and LR) with no treatment (control) and after the treatment with glibenclamide and 5-HD *in vivo*. C: the effect of DZ, glibenclamide and 5-HD on mKATP channel activity and phosphorylation efficiency. D: the correlation between mKATP channel activity, P/O, and physical endurance. M ± m, n = 6; *, # - $P < 0.05$ as compared to control.

HR group without pharmacological treatment exhibited higher mKATP channel activity and better endurance to physical stress as compared to LR group (Figure 6A, B). However, glibenclamide and 5-HD dramatically reduced physical endurance of the animals of both HR and LR groups (Figure 6A), and were equally effective in the blockage of mKATP channel *in vivo* (Figure 6B).

To find a mechanistic basis for observed phenomenon, we referred to the direct effects of mKATP channel opening by diazoxide and blocking by glibenclamide on mitochondrial functions *in vitro*. Mild uncoupling of mitochondria by diazoxide reduced the rate of phosphorylation (*J*p), but increased P/O (Figure 6D). Glibenclamide inhibited both respiration and phosphorylation, and dramatically reduced P/O (Figure 6D); J_p too was dramatically lowered by glibenclamide from 154 ± 6 to 73 ± 8 nmol·min^{-1}·mg^{-1}. So, physical endurance strongly correlated with phosphorylation efficiency, and both were dependent on mKATP channels activity (Figure 6C, D). Extreme reduction of P/O by mKATP channels blockers indicated dramatic increase of energy expense for ATP synthesis, which correlated with extreme collapse of physical endurance.

As it was shown in the literature, adaptive responses to physical stress, ischemic and hypoxic conditions trigger complex network of pathways involving the activation of protein kinases (PI3K, MAPK, PKG, PKC, etc.), NO synthase (eNOS), the modulation of glutathione and ATP levels, and the regulation of ROS production [1-5, 29, 31]. Adaptation mechanisms converged on mKATP channels activity [1, 2, 31, 85]. Our experiments too showed a strong correlation between mKATP channels activity, the efficiency of ATP synthesis P/O, and animals' endurance under exercise training. Above we hypothesized on the biochemical mechanism of the regulation of Oxphos by mKATP channels opening. However, physiological meaning of this phenomenon needs to be discussed separately. So, what could be physiological relevance of the modulation of Oxphos system by mKATP channels opening, and why the inhibition of phosphorylation could improve physical endurance in a living organism?

To answer this question, it needs to be considered that aerobic ATP synthesis is principal oxygen consuming cellular process. Thus, by our hypothesis, suppression of ATP production could prevent excess oxygen consumption by the tissues under the conditions of oxygen deficit (hypoxia, exercise training) and reduce the energy expense on phosphorylation. Contrary to this, mKATP channel blocker severely reduced P/O and extremely increased oxygen consumption under phosphorylation conditions. So, mKATP channels blocker made ATP synthesis much more energy expensive and highly oxygen consuming process. These considerations, possibly, could be directly applied to the fast exhaustion observed in the animals treated with glibenclamide and 5-HD (Figure 6A).

Thus, we have shown that bioenergetics effects of ATP-sensitive K^+ transport caused by the mild uncoupling of the respiratory chain were directly transformed into functional ones: the blocking of mPTP functioning *in vitro* and the elevation of physical endurance *in vivo*. Possibly, there was a correlation between mPTP blockage and elevated endurance of animals with higher mKATP channel activity. As it was shown, animals with higher physical endurance exhibited higher resistance to mPTP opening [83]. So, mKATP activation *in vivo* could at once produce complex effect on physical endurance: suppress mPTP opening by the reduction of ROS production, and up-regulate ATP synthesis by improving its efficiency.

Conclusion

In this work we have shown that functional effects of mKATP channels opening (the activation of K^+ cycle) directly resulted in bioenergetics effects of ATP-sensitive K^+ transport (increase in oxygen consumption and mitochondrial uncoupling). We limited our discussion of direct bioenergetics effects of ATP-sensitive K^+ transport on Ca^{2+} transport, ROS production, and ATP synthesis to the effects of 'mild uncoupling,' which in liver mitochondria occurred without depolarization. As we have shown, functional consequences of these bioenergetics effects were the inhibition of mPTP activity *in vitro* and the modulation of the Oxphos system *in vivo*.

Oxphos is one of the primary targets of mKATP channel opening *in vitro* and *in vivo*. So, as the extension to physiological conditions, we have shown physiological relevance of bioenergetics effects of ATP-sensitive K^+ transport regarding Oxphos system. MKATP channels opening increased phosphorylation efficiency and reduced oxygen consumption by Oxphos, which helped to avoid oxygen depletion under the conditions of oxygen shortage in animals subject to exhaustive exercise training. Our results provided an explanation for the improvement of the endurance in the animals with elevated endogenous mKATP channel activity.

Possible multiplicity of mKATP channels and their physiological relevance remain most important unresolved issue now. As we have shown by pharmacological means, in mitochondria, it was reasonable to speak either about one prevailing type of 'mKATP channel,' or possibly several types of KATP channels exhibiting similar properties and similar bioenergetics effects. After the novel discoveries of mKATP channels, there is a chance to resolve the problem of molecular nature of ATP-sensitive K^+ transport in mitochondria by using complex genetic, biochemical, and pharmacological approaches for the purpose of adequate understanding of molecular basis of cytoprotective effects afforded by the functioning of mKATP channels and their physiological relevance.

References

[1] Garlid KD, Costa AD, Quinlan CL, Pierre SV, Dos Santos P (2009) Cardioprotective signaling to mitochondria. *J Mol Cell Cardiol* 46: 858- 866.

[2] Pereira O. Jr., Kowaltowski A. J. Mitochondrial K^+ transport: modulation and functional consequences. *Molecules*. 2021; 26(10): 2935.

[3] Kulawiak B, Bednarczyk P, Szewczyk A. Multidimensional Regulation of Cardiac Mitochondrial Potassium Channels. *Cells.* 2021 19;10(6):1554. doi: 10.3390/cells10061554.

[4] Bláhová K., Bébarová M. ATP-sensitive potassium channels: key players in pathophysiology of many diseases. *Cas Lek Cesk*. 2021 Spring;160(2-3):71-80.

[5] Rottenberg H, Hoek JB. The path from mitochondrial ROS to aging runs through the mitochondrial permeability transition pore. *Aging Cell.* 2017; 16(5):943-955. doi: 10.1111/acel.12650.

[6] Bland A. R., Payne F. M., Ashton J. C., Jamialahmadi T., Sahebkar A. The cardioprotective actions of statins in targeting mitochondrial dysfunction associated with myocardial ischaemia-reperfusion injury. *Pharmacol Res.* 2022;175:105986. doi: 10.1016/j.phrs.2021.105986.

[7] Boovarahan S. R., Kurian G. A. Preconditioning the rat heart with 5-azacytidine attenuates myocardial ischemia/reperfusion injury via PI3K/GSK3β and mitochondrial K_{ATP} signaling axis. *J Biochem Mol Toxicol.* 2021;35(12):e22911. doi: 10.1002/jbt.22911.

[8] Aggarwal S., Randhawa P. K., Singh N., Jaggi A. S. Role of ATP-Sensitive Potassium Channels in Remote Ischemic Preconditioning Induced Tissue Protection. *J Cardiovasc Pharmacol Ther*. 2017;22(5):467-475.

[9] Peng K., Hu J., Xiao J., Dan G., Yang L., Ye F., Zou Z., Cao J., Sai Y. Mitochondrial ATP-sensitive potassium channel regulates mitochondrial dynamics to participate in neurodegeneration of Parkinson's disease. *Biochim Biophys Acta Mol Basis Dis.* 2018;1864(4 Pt A):1086-1103. doi: 10.1016/j.bbadis.2018.01.013.

[10] Nikbakht F., Khanizadeh A. M, Golab F., Baluchnejadmojarad T., Vazifehkhah S., Moeinsadat A. Mitochondrial ATP-sensitive potassium channel, MitoKATP, ameliorates mitochondrial dynamic disturbance induced by temporal lobe epilepsy. *J Chem Neuroanat*. 2021 Apr;113:101808. doi: 10.1016/j.jchemneu.2020.101808.

[11] Song N., Fang Y., Zhu H., Liu J., Jiang S., Sun S., Xu R., Ding J., Hu G., Lu M. Kir6.2 is essential to maintain neurite features by modulating PM20D1-reduced mitochondrial ATP generation. *Redox Biol.* 2021;47:102168. doi: 10.1016/j.redox. 2021.102168.

[12] Guo W., Tang Z. Y., Cai Z. Y., Zhao W. E., Yang J., Wang X. P., Ji J., Huang X. X., Sun X. L. Iptakalim alleviates synaptic damages via targeting mitochondrial ATP-sensitive potassium channel in depression. *FASEB J.* 2021;35(5):e21581. doi: 10.1096/fj.202100124RR.

[13] Jiang X., Wu D., Jiang Z., Ling W., Qian G. Protective Effect of Nicorandil on Cardiac Microvascular Injury: Role of Mitochondrial Integrity. *Oxid Med Cell Longev.* 2021 Jul 3;2021:4665632. doi: 10.1155/2021/4665632.

[14] Hu Z. L., Sun T., Lu M., Ding J. H., Du R. H., Hu G. Kir6.1/K-ATP channel on astrocytes protects against dopaminergic neurodegeneration in the MPTP mouse model of Parkinson's disease via promoting mitophagy. *Brain Behav Immun.* 2019;81:509-522. doi: 10.1016/j.bbi.2019.07.009.

[15] Liang W., Chen M., Zheng D., Li J., Song M. et al. (2017) The Opening of ATP-Sensitive K^+ Channels Protects H9c2 Cardiac Cells Against the High Glucose-Induced Injury and Inflammation by Inhibiting the ROS-TLR4-Necroptosis Pathway. *Cell Physiol Biochem.* 41:1020-1034.

[16] Koh W. U., Kim J., Lee J., Song G-W., Hwang G. S., Tak E., Song J.-G. Remote Ischemic Preconditioning and Diazoxide Protect from Hepatic Ischemic Reperfusion Injury by Inhibiting HMGB1-Induced TLR4/MyD88/NF-κB Signaling. *Int J Mol Sci*. 2019;20(23):5899. doi:10.3390/ijms20235899.

[17] Lucas A. M. B., de Lacerda Alexandre J. V., Araújo M. T. S., David C. E. B., Ponte Viana Y. I., Coelho B. N., Caldas F. R. L., Varela A. L. N., Kowaltowski A. J., Facundo H. T. Diazoxide Modulates Cardiac Hypertrophy by Targeting H_2O_2 Generation and Mitochondrial Superoxide Dismutase Activity. *Curr. Mol. Pharmacol.* 2020;13:76–83. doi: 10.2174/1874467212666190723144006.

[18] Bai S., Wang X., Wu H., Chen T., Li X., Zhang L., Li X., Er L., Du R. Cardioprotective effect of anisodamine against ischemia/reperfusion injury through the mitochondrial ATP-sensitive potassium channel. *Eur J Pharmacol.* 2021 Jun 15;901:174095. doi: 10.1016/j.ejphar.2021.174095.

[19] Zhang C. S., Han Q., Song Z. W., Jia H. Y., Shao T. P., Chen Y. P. Hydrogen gas post-conditioning attenuates early neuronal pyroptosis in a rat model of subarachnoid hemorrhage through the mitoK(ATP) signaling pathway. *Exp Ther Med.* 2021;22(2):836. doi: 10.3892/etm.2021.10268.

[20] Akopova O., Nosar V., Gavenauskas B., Bratus L., Kolchinskaya L., Mankovska I., Sagach V. (2016) The effect of ATP-dependent potassium uptake on mitochondrial functions under acute hypoxia. *J Bioenerg Biomembr.* 48(1):67-75.

[21] Mironova G. D., Shigaeva M. I., Gritsenko E. N., Murzaeva SV, Gorbacheva OS, Germanova EL, Lukyanova LD. Functioning of the mitochondrial ATP-dependent potassium channel in rats varying in their resistance to hypoxia. Involvement of the channel in the process of animal's adaptation to hypoxia. *J Bioenerg Biomembr.* 2010;42:473-481.

[22] Tregub P. P., Malinovskaya N. A., Osipova E. D., Kuzovkov D. A. Hypercapnia Modulates the Activity of Adenosine A1 Receptors and mitoK+ATP-Channels in Rat Brain When Exposed to Intermittent Hypoxia. *NeuroMolecular Medicine* 2021. doi: 10.1007/s12017-021-08672-0.

[23] Chen Xi Y., Wang J. Q., Cheng S. J., Wang Y., Deng M. Y., Yu T., Wang H. Y., Zhou W. J. Diazoxide Post-conditioning Activates the HIF-1/HRE Pathway to Induce Myocardial Protection in Hypoxic/Reoxygenated Cardiomyocytes. *Front Cardiovasc Med.* 2021;8:711465. doi: 10.3389/fcvm.2021.711465.

[24] Li J., Zhou W., Chen W., Wang H., Zhang Yu, Yu T. Mechanism of the hypoxia inducible factor 1/hypoxic response element pathway in rat myocardial ischemia/diazoxide post-conditioning. *Mol Med Rep.* 2020;21(3):1527-1536. doi: 10.3892/mmr.2020.10966.

[25] Kowaltowski A. J., Maciel E. N., Fornazari M., Castilho R. F. Diazoxide protects against methylmalonate-induced neuronal toxicity. *Exp. Neurol.* 2006;201:165–171. doi: 10.1016/j.expneurol.2006.04.004.

[26] Bai S., Wang X., Wu H., Chen T., Li X., Zhang L., Li X., Er L., Du R. Cardioprotective effect of anisodamine against ischemia/reperfusion injury through the mitochondrial ATP-sensitive potassium channel. *Eur J Pharmacol.* 2021;901:174095. doi: 10.1016/j.ejphar.2021.174095.

[27] Pecoraro M, Marzocco S, Popolo A. Diazoxide Needs Mitochondrial Connexin43 to Exert Its Cytoprotective Effect in a Cellular Model of $CoCl_2$-Induced Hypoxia. *Int J Mol Sci.* 2021;22(21):11599. doi: 10.3390/ijms222111599.

[28] Yang X., Cohen M. V., Downey J. M. Mechanism of cardioprotection by early ischemic preconditioning. *Cardiovasc Drug Ther* 2010; 24:225-234.

[29] Cohen M. V., Downey J. M. Signalling pathways and mechanisms of protection in pre- and postconditioning: historical perspective and lessons for the future. *Br J Pharmacol.* 2015; 172(8):1913-32. doi: 10.1111/bph.12903.

[30] Zi C., Zhang C., Yang Y., Ma J. Penehyclidine hydrochloride protects against anoxia/reoxygenation injury in cardiomyocytes through ATP-sensitive potassium channels, and the Akt/GSK-3beta and Akt/mTOR signaling pathways. *Cell Biol Int.* 2020;44(6):1353-1362. doi: 10.1002/cbin.11329.

[31] Daiber A. (2010) Redox signaling (cross-talk) from and to mitochondria involves mitochondrial pores and reactive oxygen species. *Biochimica et Biophysica Acta* 1797: 897-906.

[32] Yang D. M., Huang C. C., Chang Y. F. Combinatorial roles of mitochondria and cGMP/PKG pathway in the generation of neuronal free Zn^{2+} under the presence of nitric oxide. *J Chin Med Assoc.* 2020;83(4):357-366. doi: 10.1097/JCMA.000 0000000000280.

[33] Lukowski R., Cruz Santos M., Kuret A., Ruth P. cGMP and mitochondrial K^+ channels-Compartmentalized but closely connected in cardioprotection. *Br J Pharmacol.* 2021. doi: 10.1111/bph.15536.

[34] Perrelli M. G., Tullio F., Angotti C., Cerra M. C., Angelone T., Tota B., Alloatti G., Penna C, Pagliaro P. Catestatin reduces myocardial ischaemia/reperfusion injury: involvement of PI3K/Akt, PKCs, mitochondrial KATP channels and ROS signalling. *Pflugers Arch.* 2013;465(7):1031-40. doi: 10.1007/s00424-013-1217-0.

[35] Huang L., Li B., Tang S., Guo H., Li W., Huang X., Yan W., Zou F. Mitochondrial KATP Channels Control Glioma Radioresistance by Regulating ROS-Induced ERK Activation. *Mol Neurobiol.* 2015 Aug;52(1):626-37. doi: 10.1007/s12035-014-8888-1.

[36] Flagg T. P., Enkvetchakul D., Koster J. C., Nichols C. G. (2010) Muscle KATP channels: recent insights to energy sensing and myoprotection. *Physiol Rev* 90: 799-829.

[37] Raval A. P., Dave K. R., DeFazio R. A., Perez-Pinzon M. A. EpsilonPKC phosphorylates the mitochondrial K(+) (ATP) channel during induction of ischemic preconditioning in the rat hippocampus. *Brain Res.* 2007; 1184:345-53.

[38] Costa A. D. T., Jakob R., Costa C. L., Andrukhiv K., West I. C., Garlid K. D. The mechanism by which the mitochondrial ATP-sensitive K^+ channel opening and

H_2O_2 inhibit the mitochondrial permeability transition. *J. Biol. Chem.* 2006;281:20801–20808. doi: 10.1074/jbc.M600959200.

[39] Barsukevich V., Basalay M., Sanchez J., Mrochek A., Whittle J., Ackland G. L., Gourine A. V., Gourine A. Distinct cardioprotective mechanisms of immediate, early and delayed ischaemic postconditioning. *Basic Res Cardiol.* 2015; 110(1):452. doi: 10.1007/s00395-014-0452-7.

[40] Shukry M., Kamal T., Ali R., Farrag F., Almadaly E., et al. (2015) Pinacidil and levamisole prevent glutamate-induced death of hippocampal neu ronal cells through reducing ROS production. *Neurol Res* 37: 916-923.

[41] Gupta S., Sharma B. Protective effects of phosphodiesterase-1 (PDE1) and ATP sensitive potassium (KATP) channel modulators against 3-nitropropionic acid induced behavioral and biochemical toxicities in experimental Huntington's disease. *Eur J Pharmacol.* 2014;732:111-22. doi: 10.1016/j.ejphar.2014.03.032

[42] Lisa D. H., Terje H. L., Kjell O. F., Fredrik L., Jan S. Diazoxide protects against doxorubicin-induced cardiotoxicity in the rat. (2014) *BMC Pharmacology and Toxicology* 15:28.

[43] Riess M. L., Camara A. K., Heinen A., Eells J. T., Henry M. M, Stowe DF (2008) KATP channel openers have opposite effects on mitochondrial respiration under different energetic conditions. *J Cardiovasc Pharmacol* 51: 483-491.

[44] Murphy M. P. How mitochondria produce reactive oxygen species (2009) *Biochem* J. 417:1-13.

[45] Sánchez-Duarte E., Cortés-Rojo C., Sánchez-Briones L. A., Campos-García J., Saavedra-Molina A., Delgado-Enciso I., López-Lemus U.A., Montoya-Pérez R.. Nicorandil affects mitochondrial respiratory chain function by increasing Complex III activity and ROS production in skeletal muscle mitochondria. *J Membr Biol.* 2020;253(4):309-318. doi: 10.1007/s00232-020-00129-y.

[46] Jiao J. D., Garg V., Yang B., Hu K. Novel functional role of heat shock protein 90 in ATP-sensitive K^+ channel-mediated hypoxic preconditioning. *Cardiovasc Res.* 2008; 77(1):126-133.

[47] Waza A. A., Andrabi K., Hussain M. U. Protein kinase C. (PKC) mediated interaction between connexin43 (Cx43) and K(+)ATP channel subunit Kir6.1 in cardiomyocyte mitochondria: Implications in cytoprotection against hypoxia induced cell apoptosis. *Cellular Signaling*; 2014:26(9):1909-1917.

[48] Wrzosek A., Augustynek B., Żochowska M., Szewczyk A. Mitochondrial Potassium Channels as Druggable Targets. *Biomolecules.* 2020; 10(8): 1200. doi: 10.3390/biom10081200.

[49] Ardehali H., Chen Z., Ko Y., Mejía-Alvarez R., Marbán E. (2004) Multiprotein complex containing succinate dehydrogenase confers mitochondrial atp-sensitive K^+ channel activity. *Proc Natl Acad Sci* 101: 11880-11885.

[50] Foster D. B., Ho A. S., Rucker J., Garlid A. O, Chen L. et al. (2012) Mitochondrial ROMK channel is a molecular component of mitoK(ATP). *Circ Res* 111: 446-54.

[51] Paggio A., Checchetto V., Campo A., Menabò R., Di Marco G., Di Lisa F., Szabo I., Rizzuto R, De Stefani D. Identification of an ATP-sensitive potassium channel in mitochondria. *Nature.* 2019;572(7771):609-613.

[52] Juhaszova M., Kobrinsky E., Zorov D. B., Nuss H. B., Yaniv Y., Fishbein K. W., De Cabo R., Montoliu L., Gabelli S. B., Aon M. A., Cortassa S., Sollott S. J. ATP Synthase K^+- and H^+-Fluxes Drive ATP Synthesis and Enable Mitochondrial K^+-"Uniporter" Function: I. Characterization of Ion Fluxes. *Function* (Oxf) 2022; 3(2): zqab065. doi: 10.1093/function/zqab065.

[53] Foster M. N., Coetzee W. A. (2016) KATP Channels in the Cardiovascular System. *Physiol Rev* 96:177-252.

[54] Moreau C., Prost A. L., Derand R., Vivaudou M. (2005) SUR, ABC proteins targeted by KATP channel openers. *J Mol Cell Cardiol* 38: 951-963.

[55] Flagg T. P., Kurata H. T., Masia R., Caputa G., Magnuson M. A. et al. Differential structure of atrial and ventricular KATP: atrial KATP channels require SUR1. *Circ. Res.* 2008;103:1458–1465.

[56] Liu Y., Ren G., O'Rourke B., Marbán E., Seharaseyon J. Pharmacological comparison of native mitochondrial K(ATP) channels with molecularly defined surface K(ATP) channels. *Mol Pharmacol.* 2001;59:225-230.

[57] Nakagawa Y., Yoshioka M., Abe Y., Uchinami H., Ohba T. et al. Enhancement of liver regeneration by adenosine triphosphate-sensitive K^+ channel opener (diazoxide) after partial hepatectomy. *Transplantation*. 2012;93:1094-1100.

[58] Nelson P. T., Jicha G. A., Wang W. X., Ighodaro E., Artiushin S. et al. ABCC9/SUR2 in the brain: Implications for hippocampal sclerosis of aging and a potential therapeutic target. *Ageing Res Rev*. 2015;24(Pt B):111-125.

[59] Jabůrek M., Yarov-Yarovoy V., Paucek P., Garlid K. D. State-dependent inhibition of the mitochondrial K(ATP) channel by glyburide and 5-hydroxydecanoate. *J. Biol. Chem.* 1998;273:13578–13582. doi: 10.1016/S0021-9258(19)57796-1.

[60] D'hahan N., Moreau C., Prost A. L., Jacquet H., Alekseev A. E., et al. Pharmacological plasticity of cardiac ATP-sensitive potassium channels toward diazoxide revealed by ADP. *Proc Natl Acad Sci USA*. 1999;96:12162–12167.

[61] Inoue I., Nagase H., Kishi K., Higuti T. (1991) ATP-sensitive K^+ channel in the mitochondrial inner membrane. *Nature.* 352: 244-247.

[62] Mironova G. D., Negoda A. E., Marinov B. S., Paucek P., Costa A. D. T., et al. (2004) Functional distinctions between the mitochondrial ATP-depen- dent K^+ Channel (mitoK) and its inward rectifier subunit (mitoKIR). *J. Biol. Chem* 279: 32562-32568.

[63] Garlid K. D., Paucek P., Yarov-Yarovoy V., Sun X., and Schindler P. A. (1996) The mitochondrial K_{ATP} channel as a receptor for potassium channel openers. *J Biol Chem.* 271:8796–8799.

[64] Kopustinskiene D. M., Liobikas J., Skemiene K., Malinauskas F., Toleikis J. (2010) Direct effects of KATP channel openers pinacidil and diazoxide on oxidative phosphorylation of mitochondria in situ. *Cell Physiol Biochem* 25: 181-186.

[65] Akopova O. V., Nosar V. I., Bouryi V. A., Mankovska I. N., Sagach V. F. (2010) Influence of ATP-dependent K^+-channel opener on K^+-cycle and oxygen consumption in rat liver mitochondria. *Biochemistry* (Moscow) 75: 1139-1147.

[66] Akopova O. V., Kolchinskaya L. I., Nosar V. I., Bouryi V. A., Mankovska I. N., et al. (2014) Effect of potential-dependent potassium uptake on production of reactive oxygen species in rat brain mitochondria. *Biochemistry* (Mosc) 79: 44-53.

[67] Wang J., Bai J., Duan P., Wang H., Li Y., Zhu Q. Kir6.1 improves cardiac dysfunction in diabetic cardiomyopathy via the AKT-FoxO1 signalling pathway. *J Cell Mol Med*. 2021;25(8):3935-3949. doi: 10.1111/jcmm.16346.

[68] Wojtovich A. P., Urciuoli W. R., Chatterjee S., Fisher A. B., Nehrke K., et al, (2013) Kir6.2 is not the mitochondrial KATP channel but is required for cardioprotection by ischemic preconditioning. *Am J Physiol* 304: 1439-1445.

[69] Akopova O., Kolchinskaya L., Nosar V., Mankovska I., Sagach V. Diazoxide affects mitochondrial bioenergetics by the opening of mKATP channel on submicromolar scale. *BMC Mol Cell Biol*. 2020;21(1):31.

[70] Beavis A. D. (1987) Upper and lower limits of the charge translocation stoichiometry of mitochondrial electron transport. *J Biol Chem* 262:6165-6173.

[71] Garlid K. D., Paucek P (2003) Mitochondrial potassium transport: the K(+) cycle. *Biochim Biophys Acta*. 1606:23-41.

[72] Akopova O. V., Kolchinskaya L. I., Nosar V. I., Smirnov A. N., Bratus L. B. ATP-sensitive potassium transport of rat brain mitochondria is highly sensitive to mKATP channels openers: a light scattering study. *Ukr. Biochem. J.* 2020;92(5): 5-14.

[73] Kowaltowski A. J., Seetharaman S., Paucek P., Garlid K. D. Bioenergetic consequences of opening the ATP-sensitive K^+ channel of heart mitochondria. *Am. J. Physiol. Hear Circ. Physiol.* 2001;280:649–657. doi: 10.1152/ajpheart.2001.280.2.H649.

[74] Holmuhamedov E. L., Jovanović S., Dzeja P. P., Jovanović A., Terzic A. (1998) Mitochondrial ATP-sensitive K^+ channels modulate cardiac mitochondrial function. *Am J Physiol.* 275:H1567-H1576.

[75] Murata M., Akao M., O'Rourke B., Marbán E. (2001) Mitochondrial ATP-sensitive potassium channels attenuate matrix $Ca^{(2+)}$ overload during simulated ischemia and reperfusion: possible mechanism of cardioprotection. *Circ Res.* 89:891-898.

[76] Akopova O. V., Nosar V. I., Mankovskaya I. N., Sagach V. F. Calcium uptake in rat liver mitochondria accompanied by activation of ATP-dependent potassium channel. *Biochemistry (Mosc).* 2008;73(10):1146-53.

[77] Koeppe R. E., Rottenberg H. Mechanism of uncoupling of oxidative phosphorylation by gramicidin. *Biochemistry.* 1989;28:4355-4360.

[78] Stucki JW. Efficiency of oxidative phosphorylation and energy dissipation by H+ ion recycling in rat liver mitochondria metabolizing pyruvate. *Eur J Biochem.* 1976;68:551-562.

[79] Zorov D. B., Juhaszova M., Yaniv Y., Nuss H. B., Wang Su, Sollott S. J. Regulation and pharmacology of the mitochondrial permeability transition pore. *Cardiovasc Res.* 2009;83:213-225.

[80] Akopova O. V., Kolchinskaya L. I., Nosar V. I., Buryĭ V. A., Mankovska I. N., Sagach VF. [The effect of ATP-dependent K(+)-channel opener on transmembrane potassium exchange and reactive oxygen species production upon the opening of mitochondrial pore]. *Ukr Biochem J.* 2014; 86(2):26-40. (Russian).

[81] Mankovska I. N., Nosar V. I., Gorbacheva O. S., Gonchar O. A., Gavenauskas B. L., Bratus L. V., Mironova G. D. The effect of uridine on the endurance of animals

with different resistance to physical stress: the role of mitochondrial ATP-dependent potassium channel. *Biophysics*. 2014;59:764-767.

[82] Mironova G. D., Pavlik L. L., Kirova Y. I., Belosludtseva N. V., Mosentsov A. A., Khmil N. V., Germanova E. L., Lukyanova L. D. Effect of hypoxia on mitochondrial enzymes and ultrastructure in the brain cortex of rats with different tolerance to oxygen shortage. *J Bioenerg Biomembr*. 2019;51(5):329-340.

[83] Aon M. A., Cortassa S., Juhaszova M., Gonzalez-Reyes J. A., Calvo-Rubio M., Villalba J. M., Lachance A. D., Ziman B. D., Mitchell S. J., Murt K. N, Axsom J. E. C, Alfaras I., Britton S. L., Koch L. G., de Cabo R., Lakatta E. G., Sollott S. J. Mitochondrial health is enhanced in rats with higher vs. lower intrinsic exercise capacity and extended lifespan. *Aging and Mechanisms of Disease*. 2021;7:1. https://doi.org/10.1038/s41514-020-00054-3.

[84] Baranova O. V., Skarga Y. Y., Negoda A. E., Mironova G. D. (2000) Inhibition of DNP-induced potassium efflux by adenine nucleotides in mitochondria. *Biokhimia* (Moscow) 65:262-267 (*in Russian*)

[85] Murphy E. Primary and secondary signaling pathways in early preconditioning that converge on the mitochondria to produce cardioprotection. *Circ Res*. 2004;94: 7-16.

Chapter 3

Bilirubin Interactions with Different Proteins and Implications of These Interactions

Nikola Gligorijević[1,*] and Simeon Minić[2]
[1]Departemnt of Metabolism, University of Belgrade – Institute for the Application of Nuclear Energy, Belgrade, Serbia
[2]Center of Excellence for Molecular Food Sciences and Department of Biochemistry, University of Belgrade – Faculty of Chemistry, Belgrade, Serbia

Abstract

Bilirubin is a tetrapyrrole, yellow pigment, and it is a degradation product created in heme metabolism. This molecule is predominantly present in its unconjugated form in circulation, and it is mainly bound to human serum albumin. The presence of small amounts of free, unconjugated bilirubin enables its interactions with other circulation components. In conditions like Gilbert syndrome or diabetes, the concentration of unconjugated bilirubin increases in the blood, making other interacting partners from circulation more important for its metabolism. In the liver, bilirubin is converted to conjugated form, which is then excreted from the organism. While the significant increase of bilirubin is toxic, research suggests that small increases may be beneficial since bilirubin has antioxidative and anticancer potential. In this chapter, interactions of bilirubin with several proteins will be described together with the effects of these interactions on the protein's structure and function.

Keywords: bilirubin, protein binding, metabolism, phycocyanobilin

---

* Corresponding Author's Email: nikolag@inep.co.rs.

In: Advances in Biology. Volume 1
Editor: Charles D. Grant
ISBN: 979-8-88697-201-6

Metabolism of Bilirubin

Bilirubin is a tetrapyrrole molecule that is produced from heme. Its production occurs from two pathways. The first and dominant one is the breakdown of hemoglobin from erythrocytes. About 80% of bilirubin is produced like this. Other pathway includes the breakdown of other heme-containing proteins from the liver and muscles, including myoglobin, catalase, peroxidase, cytochromes and tryptophan pyrrolase. About 4 mg of bilirubin per kg of body weight is produced daily [1].

Heme, an important iron-complexing tetrapyrrole, is converted to bilirubin by the reticuloendothelial system, mainly in the liver and spleen [2]. This enzymatically catalyzed reaction has two steps, Figure 1. First, heme is specifically cleaved by hem oxygenase at the α-methene bridge. This reaction crates biliverdin, free iron and carbon monoxide. Three molecules of oxygen and NAPDH are required for this reaction. In the second step, biliverdin is reduced to bilirubin by biliverdin reductase. Heme conversion to bilirubin represents the most important endogenous production pathway of CO and the formation of bilirubin can be quantified from the exhaled CO in breath [1].

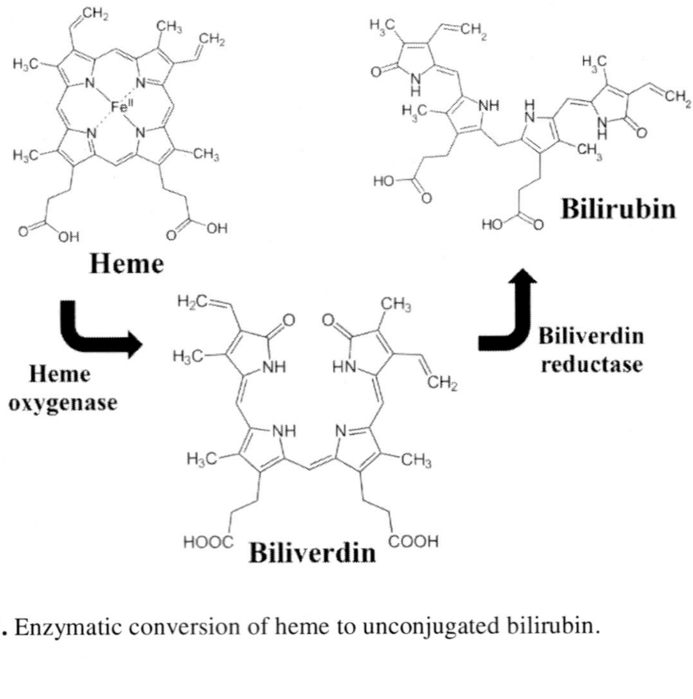

Figure 1. Enzymatic conversion of heme to unconjugated bilirubin.

The dominant bilirubin isomer in human serum is IX-alpha (4Z,15Z). Internal hydrogen bonds between IX-alpha (4Z,15Z) bilirubin side groups make this isomer hydrophobic. This enables bilirubin to cross phospholipid membranes, but also in fetal life bilirubin can cross the placenta and enter the maternal organism for excretion. During the postnatal period, bilirubin can pass across the blood-brain barrier, making jaundice in newborns potentially serious condition [3].

Created bilirubin is also called unconjugated. It enters circulation tightly bound to its main carrier, serum albumin. Since unconjugated bilirubin is not water-soluble, it cannot be excreted in urine or bile. Formed complexes of bilirubin and albumin enter the liver. At the sinusoidal surface of hepatocytes, the formed complex dissociates. Hepatocytes take bilirubin with the energy-consuming transport that requires Cl^- and Cl^-/HCO_3^- exchange. Inside, cytosolic glutathione-S-transferases, also called ligandins, bind unconjugated bilirubin, thus keeping it soluble and lowering its leaving from the cell.

Glucuronidation of bilirubin is very important for its excretion via the biliary tract. A specific isoform of uridinediphosphoglucuronate glucuronosyltransferase, named UGT1A1, catalyzes this reaction [4]. Conjugated, water-soluble bilirubin is transported into the bile canaliculus against a concentration gradient by utilizing an energy-consuming mechanism. The energy is derived by ATP-hydrolysis performed by a canalicular membrane protein named ABCC2. This protein is also called multidrug resistance-related protein-2 (MRP2) [5]. The same pathway is used to excrete many organic anions, especially those conjugated with glutathione and glucuronic acid. From the biliary tract, bilirubin reaches the intestine. Conjugated bilirubin is not reabsorbed here and its water-solubility contributes to this [6]. Bilirubin monoglucuronides and diglucuronides hydrolyze back into unconjugated bilirubin. Microbial and intestinal mucosa β-glucuronidase facilitates this process [7]. About 25% of again formed unconjugated bilirubin is reabsorbed by the intestine and returns to enterohepatic circulation. The remainder, together with conjugated bilirubin, is reduced by anaerobic gut microbiota and three colorless tetrapyrroles are formed, named stercobilinogen, urobilinogen, or mesobilinogen, generally known as urobilinogens. These molecules spontaneously oxidize to yellow-brown bile pigments, stercobilin, urobilin and mesobilin. Up to one-fifth of daily produced urobilinogens are reabsorbed by the intestine and re-enter enterohepatic circulation. A dominant fraction of this urobilinogen is again excreted into bile, while a small amount enters general circulation is removed by urine [6]. The pathway of bilirubin is presented in Figure 2.

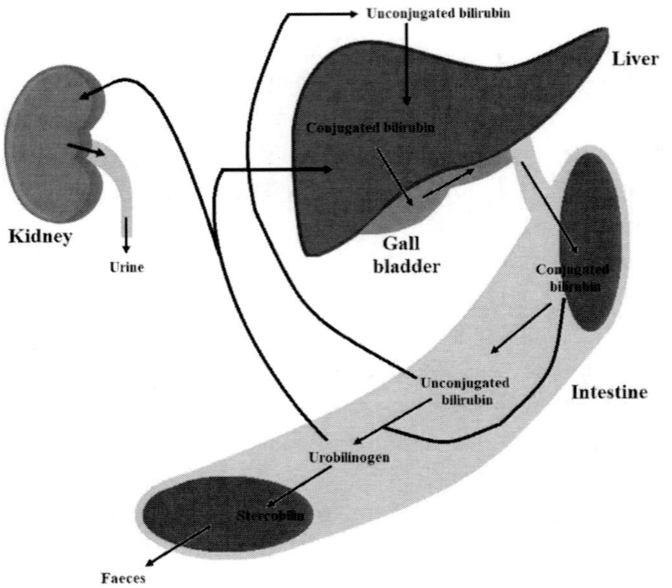

Figure 2. Pathways of bilirubin excretion.

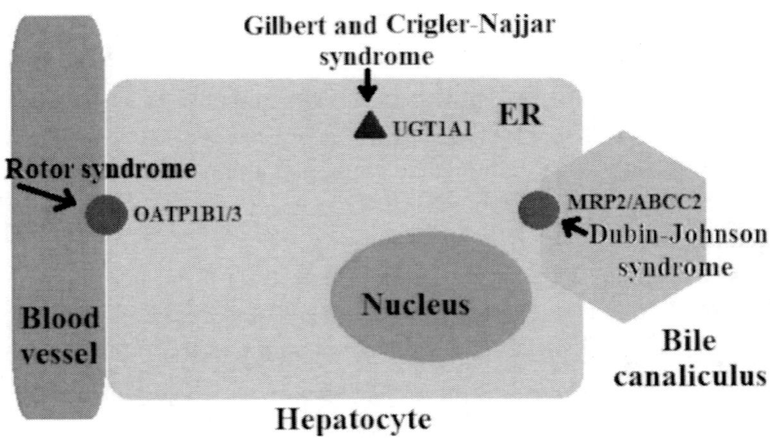

Figure 3. Location of different proteins and disorders of bilirubin metabolism associated with them. ER- smooth endoplasmic reticulum, OATP1B1/3- organic anion-transporting proteins, UGT1A1- uridinediphosphoglucuronate glucuronosyltransferase, MRP2/ABCC2- multidrug resistance-associated protein 2/adenosine triphosphate (ATP)-binding cassette subfamily C member 2.

Health Implications of Bilirubin

At many stages of bilirubin metabolism, defects can occur, leading to abnormalities in bilirubin excretion and increased concentration. While a high bilirubin concentration is toxic, some newer data suggest that marginal increases of bilirubin could have protective effects. Some of the syndromes, Figure 3, characterized by increased bilirubin, will be described in this part as well as health implications that may arise as a consequence of these conditions.

Metabolic Syndromes That Cause Hyperbilirubinemias

Organic anion-transporting proteins (OATP1B1 and OATP1B3) are membrane transporters involved in liver uptake of many molecules, including conjugated and unconjugated bilirubin. The absence of these proteins in the liver causes a disorder known as Rotor syndrome (RS) [8]. RS is an autosomal recessive disorder with mild, mainly conjugated hyperbili-rubinemia. Other than jaundice, if apparent, no other symptoms exist and clinical results are normal. Liver biopsy is also normal. The condition does not require any treatment and the prognosis is excellent [9]. The mutations that lead to a deficiency of OATP1B1 and OATP1B3 affect genes that encode them, *SLCO1B1* and *SLCO1B3*. Interestingly, both genes need to be affected to develop Rotor syndrome, which may explain the rarity of this condition [10]. Although no related drug toxicity in persons with RS has been reported so far, it is known that drugs like statins, anticancer agents and methotrexate accumulate at higher concentrations in tissues and blood [11, 12]. This is because OATP1B1 is involved in drug detoxification and these results should be taken with caution to avoid potential toxicity.

UGT1A1 is the only enzyme that glucuronidates bilirubin and mutations that entirely prevent its activity lead to the total absence of conjugated bilirubin [13]. UGT1A1 is a transmembrane protein, mainly present on the smooth endoplasmic reticulum. On its luminal side, UGT1A1 has places for binding of both bilirubin and glucuronic acid. In the lumen of the endoplasmic reticulum, UGT1A1 catalyzes the formation of mono- and diglucuronides. Formed conjugated bilirubin then leaves endoplasmic reticulum, enters the cytoplasm of hepatocyte and is being excreted via the bile. Syndromes associated with the mutations of the *UGT1A1* gene are known as Gilbert and Crigler–Najjar Syndrome [9].

Gilbert syndrome (GS) is a benign autosomal recessive disorder characterized by mild unconjugated hyperbilirubinemia. Liver function in this condition is normal. Bilirubin levels in patients usually do not exceed 70 μM. Serum bilirubin levels are increased during fasting. This condition typically does not require treatment with medications, but if needed, phenobarbital is prescribed, which reduces bilirubin concentration to normal levels. On a genetic level, the most common defect of Gilbert syndrome is the addition of extra TA dinucleotide to the TATA promoter sequence of the *UGT1A1* gene [14]. In persons homozygous for this mutation, the activity of UGT1A1 is about 10 to 35% of normal since the synthesis of a functional enzyme is reduced. In bile, bilirubin is dominantly diglucuronidated, but the proportion of monoglucuronide is larger when compared to an individual without the syndrome [9]. For hyperbilirubinemia to develop, a person needs to be homozygous for a mutation in the promotor sequence. However, this is not enough. Some other disorder(s) should also be present, like mild hemolysis, defects in bilirubin uptake or other mutations [15, 16]. If a patient has GS, the metabolism of some drugs can be affected, which is much more serious than GS itself. Molecules that are metabolized in hepatocytes by glucuronidation cannot be processed in GS efficiently. Some of them are menthol, estradiol benzoate, lamotrigine, tolbutamide, ethinyl estradiol, statins, acetaminophen etc. [17, 18]. Protease inhibitors of the human immunodeficiency virus (HIV), indinavir and atazanavir, inhibit the activity of UGT1A1. Hyperbilirubinemia is more pronounced if the person receiving these protease inhibitors has GS as well, however rarely significant toxicity occurs [19]. Antitumor agent irinotecan is so far the main danger for serious drug-induced toxicity in individuals with GS whose active substance needs to be glucuronidated by UGT1A1 [20, 21]. Toxic effects include myelosuppression with severe neutropenia and intractable diarrhea.

Crigler–Najjar syndrome (CN) is an autosomal recessive disorder that is characterized by significant unconjugated hyperbilirubinemia in a concentration range from 100 to 750 μM. CN occurs due to mutations in the *UGT1A1* gene. Two types of this syndrome exist. Type I is characterized by almost complete abolishment of UGT1A1 activity with the complete absence of conjugated bilirubin. Concentrations of unconjugated bilirubin may reach 750 μM [9]. Treatment of patients with phenobarbital does not improve the condition. Type I of CN represents a severe condition and persons have a high risk for developing bilirubin encephalopathy. If the neonatal period is survived, children remain at high risk for developing later bilirubin encephalopathy [22]. In this type of CN mutations are found in exons 1 to 5.

If exon 1 is affected, then only glucuronidation of bilirubin is affected. On the other hand, if mutations are on other exons, then glucuronidation of other substrates is heavily affected. Mutations result in either premature truncation of protein or substitution of important amino acid [23]. Liver transplantation is the best option for the prevention of brain injury and death and should be performed as soon as possible [24]. Phototherapy is also applied, but it is not always effective. Some studies suggest that gene therapy could potentially treat CN type I [25]. Compared to CN type I, CN type II is a milder condition where the activity of UGT1A1 is partially maintained (about 10%, compared to normal). Concentrations of bilirubin reach up to 400 μM. In bile bilirubin monoglucuronide is dominant [26]. Treatment with phenobarbital reduces bilirubin by about 25%. Treatment of CN type II with phenobarbital is lifelong and it usually prevents bilirubin encephalopathy. A single amino acid residue substitution in UGT1A1 is responsible for this syndrome. This substitution greatly reduces, but not completely, the activity of UGT1A1 [23].

Dubin and Johnson initially described in 1954 a benign syndrome characterized by predominantly conjugated hyperbilirubinemia, without hemolysis. This condition was named Dubin-Johnson syndrome (DJS) and it is an autosomal recessive disorder caused by genetic variations of adenosine triphosphate (ATP)-binding cassette subfamily C member 2 (*ABCC2*) gene. Patients with DJS may have nonspecific symptoms, for example, abdominal discomfort [27]. Some other symptoms may include fatigue and fever. Sometimes, however rarely, bilirubin levels can be so high to cause organ damage. Urinary excretion of coproporphyrin is normal, with coproporphyrin I representing the majority of the coproporphyrin fraction [28]. Liver histology from DJS patients shows that these patients have black-colored livers. The color originates from the deposition of melanin-like pigment in lysosomes of hepatocytes. This disorder that occurs among all races, nationalities, and both sexes is considered a rare disorder. It is, however, more present in Jews [29]. Diagnosis of Dubin-Johnson syndrome is rare in the neonatal period. Still, if the syndrome presents itself with severe cholestasis at that age, then the term neonatal Dubin-Johnson syndrome is used [30]. *ABCC2* gene is located on chromosome 10q24 and encodes the human canalicular multispecific organic anion transporter, also known as the multidrug resistance-associated protein 2 (MRP2). This protein is a non-bile acid organic anion transporter. It is involved in the primary active export of lipophilic compounds conjugates with glucuronate or glutathione from cells and it requires ATP for its activity [31]. Several genetic alterations of the *ABCC2* gene, including deletions, missense mutations, nonsense mutations and splice junction mutations, have been identified in DJS patients [32, 33]. Mutations of the *ABCC2* gene cause defects of MRP2 protein. Some lead to fast degradation of its mRNA, while others affect protein maturation, stability and function. Whatever the case is, the main consequence is a significant reduction of biliary secretion of organic anions. Certain mutations may lead to rapid degradation of the mutant mRNA, whereas others may affect protein maturation, protein stability, or the function of correctly localized proteins [34, 35].

Jaundice in newborns is a common condition of increased bilirubin concentration. Although difficult, it is very important to distinguish physiologic jaundice from pathologic one in newborns [3]. Physiologic jaundice in newborns normally occurs due to increased breakdown of red cells with the presence of a lower capacity for uptake, conjugation, and excretion of bilirubin by the liver. Some of these factors that produce hyperbilirubinemia can be amplified, thus leading to more severe forms of neonatal jaundice. The fetus's environment is poor in oxygen and for this reason, a fetus has structurally different hemoglobin from the one in postnatal life. This fetus hemoglobin is also present at higher concentrations. After birth, there is no need for this low oxygen compensation and the excess of hemoglobin is degraded. Due to some immunity disorders, the presence of toxins and certain drugs, or congenital hemolytic anemias, increased breakdown of hemoglobin may occur in newborns, leading to more severe jaundice. Neonates do not have ligandin, a protein that binds bilirubin in hepatocytes. While this may contribute to increased bilirubin, there is no proof that deficiency of this protein is involved in pathologic jaundice. Increased enterohepatic circulation may lead to jaundice in newborns as well. It happens due to several factors like the lack of oral feeding due to some illness or the lack of adequate nutrition due to the lack of breast milk. After establishing proper oral feeding and by using human milk substitutes, enterohepatic circulation is reduced [3].

Beneficial Effects of Mildly Increased Bilirubin

Mild increases of bilirubin are considered to have beneficial effects, mainly due to its antioxidative potential [36]. For example, epidemiological studies have demonstrated benefits of elevated unconjugated bilirubin, like in Gilbert syndrome [37, 38]. Bilirubin may also offer protection from cardiovascular diseases (CVD) by reduction of risk factors that lead to CVD. Bilirubin may protect both lipids and transmembrane proteins from harmful reactive oxidative species, thus reducing hyper-reactivity of platelets and stopping prothrombotic phenotype [39]. Lower LDL and VLDL cholesterol, together with an increased ratio of HDL/LDL, are associated with a higher concentration of bilirubin in serum. Unconjugated bilirubin is also involved in the improvement of diabetes and metabolic syndrome [39–41]. Reduction of cytokine release, like INFγ, TNFα and IL2, reduced complement activation and the response of T cells are all associated with unconjugated bilirubin [42]. Inhibitions of NADPH, VCAM-1 and iNOS are mechanisms of anti-inflammatory effects of unconjugated bilirubin [43–45]. Reduction of CVD risk by mild hyperbilirubinemia also comes from interfering with the vasoconstrictor effect of angiotensin II, which consequently prevents the drop of glomerular filtration rate and increases renal blood flow and systolic blood pressure [46]. Unconjugated bilirubin also prevents angiotensin II-induced expression of endothelin, which has a vasoconstrictive effect [47]. General improvement of hemodynamics in kidneys might be one possible mechanism of prevention of renal diseases by bilirubin [46]. The expression of heme oxygenase is induced by angiotensin II, which may promote the intracellular synthesis of unconjugated bilirubin [48]. Improved cell survival was reported for several cell lines like breast, colon, liver, etc., due to *de novo* synthesis of unconjugated bilirubin [49]. Increased bilirubin concentration in serum is also associated with a lower frequency of colorectal cancer [50]. By using the HRT-18 colorectal cancer cell line, it was shown that bilirubin had an anti-proliferative effect *in vivo* and a cytostatic/pro-apoptotic effect *in vitro*. This effect was dependent on mitogen-activated protein kinases (MAPK) [51].

On the other hand, lower concentrations of total serum bilirubin are associated with increased mortality due to cancer and cardiovascular complications. Several other conditions are also characterized by low bilirubin concentration. They include schizophrenia, dementia, multiple sclerosis, cerebral infarction, Alzheimer's disease, etc. Whether lower bilirubin concentration is a predisposing factor for the development of mentioned pathologies remains to be seen [46].

Interaction of Bilirubin with Proteins from Circulation

Interaction with Human Serum Albumin (HSA)

Human serum albumin is a single polypeptide globular protein composed of three homologous α-helical domains (I, II and III), each containing two subdomains (A and B). It is the most abundant plasma protein, and it has an essential role in the transportation of endogenous and exogenous ligands, including fatty acids, bilirubin, metal ions and drugs, thereby increasing their solubility and enabling them to be transported through the bloodstream [52].

HSA potently binds 4Z,15Z-isomer of bilirubin as it prevents the transfer of bilirubin from blood to tissues and impairs the development of bilirubin encephalopathy, a severe pathological condition [53, 54]. It is supposed that HSA has one primary site for bilirubin binding with the affinity constant between 10^7 and 10^8 M^{-1} [55], but there are also secondary sites with at least tenfold lower affinity [56]. However, it is still debated about the localization of bilirubin primary binding site within HSA. The two main candidates for high-affinity bilirubin binding are subdomains IB and IIA. Crystal structure of HSA complex with bilirubin photoisomer 4Z,15E revealed binding to subdomain IB [57]. However, the conformation of bilirubin isomer in this complex has M-type helicity, which is not in accordance with circular dichroism study of HSA-bilirubin complex in solution, which indicates binding of bilirubin P conformer [54]. Additionally, CD spectroscopy was utilized to monitor bilirubin's displacements from HSA binding sit(s) by site-specific ligands. This approach revealed that the high-affinity binding site for bilirubin is located at subdomain IIA [58]. Another study dealing with the phage-displayed HSA mutants demonstrated that the binding affinity of 4Z,15Z-bilirubin for wild-type domain I (residues 1–186) was significantly lower than that for wild-type domain II. On the other hand, Petersen and colleagues showed that mutations of particular residues in subdomain IIA did not significantly influence the binding of bilirubin to HSA. Based on these results, they proposed the dynamic binding mode of bilirubin to HSA [53]. This mode is attributable to the formation of a flexible 4Z,15Z-bilirubin– HSA complex, and no single crucial binding mode exists, but multiple interactions are ongoing between 4Z,15Z-bilirubin and HSA [53, 54].

Our previous study combining computational (molecular docking) and experimental approaches (fluorescence quenching) demonstrated the ability of HSA to bind bilirubin at both subdomains (IB and IIA) with a similar affinity (Figure 4) [54]. However, due to allosteric coupling between different subdomains in HSA, the binding of a ligand at subdomain IIA could influence the binding at subdomain IB and vice versa [59, 60]. Therefore, additional efforts are needed to resolve the HSA-bilirubin puzzle convincingly, and this research topic is still of high interest for further investigations.

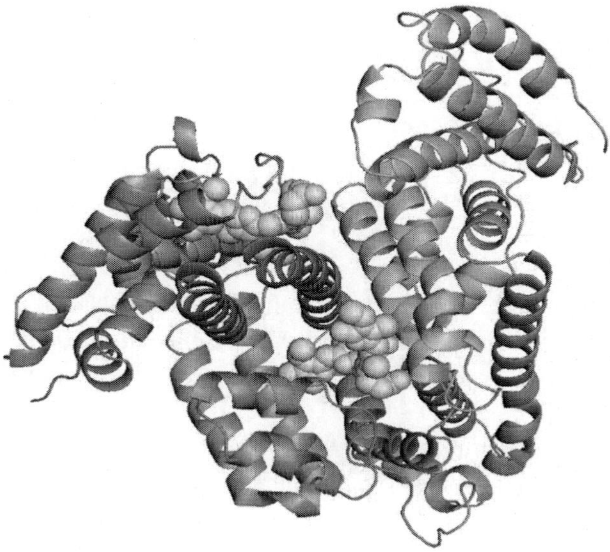

Figure 4. Molecular modeling of bilirubin binding to HSA. The domains in the ribbon model of protein HSA (PDB ID: 1BM0) are colour codes: I, orange; II, green; III, turquoise blue. Structural models depicting two potential binding sites for bilirubin to HSA: bound bilirubin is shown in yellow colour space-filling representation. Reprinted from [54].

Fatty acids represent the additional endogenous ligands with a high propensity for binding to HSA. HSA can simultaneously bind seven long-chain fatty acids (FAs) at the sites characterized as fatty acid-binding sites 1 to 7 (FABSs 1–7), which could overlap with the main drug or endogenous ligands binding sites. However, FABSs show variable affinities towards FAs, while FA7 at subdomain IIA weakly bind fatty acids ligands and the high-affinity ligands such as bilirubin could easily displace FA ligands. Moreover, it was shown that binding of FA to HSA could induce conformation changes in HSA, resulting in the higher binding affinity of bilirubin to HSA in comparison to defatted albumin. However, if the FA:albumin molar ratio is >4, displacing of bilirubin from a primary albumin-binding site occurs [61, 62]. Therefore, fluctuations in serum-free fatty acid levels could significantly modulate the bilirubin free serum level and its body distribution [63].

Impaired bilirubin binding to HSA could induce unconjugated hyperbilirubinemia, which causes bilirubin-induced neurologic dysfunction. This phenomenon is triggered by reduced HSA concentration or decreased affinity of bilirubin to HSA [61]. Decreased affinity could be induced by increased concentration of displacing molecules such as fatty acids or drugs [61, 64]. Furthermore, various pathophysiological conditions such as infection, hypothermia, hypoxia, acidosis, hypercarbia and asphyxia could also impair the binding of bilirubin to HSA [61].

Alpha-fetoprotein

Alpha-fetoprotein (AFP) is a major mammalian embryo-specific and tumour-associated protein. It is a glycoprotein with a molecular weight in the range 68- 73 kDa [65]. AFP is a product of albuminoid genes, and hence, it has a high structure similar to HSA [65], indicating its ability to bind various ligands, including bilirubin. Indeed, it was demonstrated that AFP binds bilirubin with high affinity [65, 66], and the location of binding sites is similar to HSA. Therefore, AFP could serve as a carrier protein in fetal serum, confirming its functional homology with HSA. However, the high-affinity binding site of AFP binds bilirubin with a lower affinity (10^6-10^7 M^{-1}) in comparison to the HSA-bilirubin system [66, 67]. AFP could also bind bilirubin at a relatively low-affinity site with a binding constant ~ 105 M^{-1} [67].

Apolipoprotein D

Apolipoprotein D (apo D) is a 29-kDa glycoprotein primarily associated with high-density lipoproteins (HDL) in human plasma. It is a member of the lipocalin family, which adopts a β-barrel tertiary structure and transport small hydrophobic ligands [68, 69]. Interestingly, tetrapyrrole-based ligands are more favourable for apo D than lipid molecules such as cholesterol or cholesteryl ester [70]. Indeed, Goessling and Zucker demonstrated high-affinity (~10^7 M^{-1}) bilirubin binding to human apoD with stoichiometry 1:1. Additionally, incorporating apo D into phosphatidylcholine vesicles did not affect the binding constant, suggesting that a lipid environment does not alter the affinity of the protein for bilirubin. The same study performed the kinetic characterizations of Apo D-bilirubin and HDL-bilirubin complexes. A double exponential function best describes bilirubin dissociation from HDL: the fast component of dissociation from HDL represents bilirubin solvation from surface phospholipids, while the slow component is related to the bilirubin dissociation from apo D. However, both dissociation rate constants are significantly higher than the dissociation rate constant for HSA-bilirubin system. Therefore, dissociation of bilirubin from HSA is the rate-limiting step in the hepatocellular uptake of this metabolite. On the other hand, bilirubin dissociation from HDL (and apo D) is significantly faster, which makes HDL (or Apo D) a more proficient bilirubin donor [71].

The risk of neonatal bilirubin-induced neurotoxicity correlates most closely with the concentration of nonalbumin-bound bilirubin in the plasma, and the HDL is the principal nonalbumin carrier of bilirubin. Therefore, HDL-bound bilirubin is an important mediator of bilirubin toxicity, probably due to increased availability for uptake into the central nervous system [71, 72].

Alpha-2-Macroglobulin

Alpha-2-macroglobulin (α2M) is a 720 kDa homotetrameric circulatory glycoprotein, which serves as a universal proteinase inhibitor [73]. Combining the multi-spectroscopic approach with an isothermal titration calorimetry study demonstrated exothermic, moderate-affinity binding (~10^4 M^{-1}) of bilirubin to the α2M. Additionally, the presence of bilirubin at physiological concentrations induces significant loss of α2M anti-proteolytic potential, which could be described by bilirubin-induced conformational changes of α2M through the transition of the β-sheet to α-helix [74].

Interaction with Fibrinogen

Fibrinogen, or factor I of the coagulation system, is a fibrillar protein whose dominant form in circulation has a molecular mass of 340 kDa. Several other types of fibrinogen exist, but they are in the minority. Their occurrence is a consequence of the alternative splicing of mRNA. Fibrinogen with a larger Aα-chain is called fibrinogen-420 or α_EC, and the one with larger γ-chain is called γ'-fibrinogen. The primary function of this protein is in hemostasis. Fibrinogen is involved in both primary and secondary hemostasis. Considering the susceptibility of fibrinogen towards oxidation and antioxidant abilities of bilirubin, the interaction between these two molecules under physiological conditions was characterized, together with the consequences of this interaction on the structure and stability of fibrinogen, fibrin formation and protection against oxidation [75].

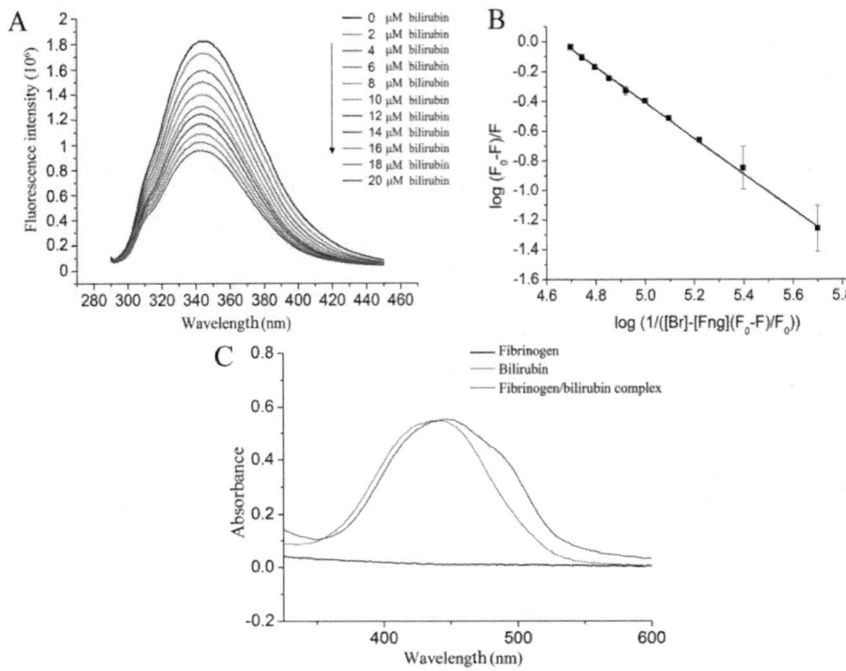

Figure 5. The emission spectra of fibrinogen in the absence and in the presence of different concentrations of bilirubin (A). Determination of the binding constant of the fibrinogen/bilirubin complex (B). UV-VIS spectra of bilirubin, fibrinogen and bilirubin/fibrinogen complex (C). Reprinted from [75].

Binding affinity between fibrinogen and bilirubin (unconjugated bilirubin was used) was determined using spectrofluorimetry. Bilirubin caused the concentration-dependent quenching of intrinsic fluorescence of fibrinogen, Figure 5A. From this quenching data, the affinity constant was calculated to be 4.5×10^4 M^{-1}, Figure 5B, suggesting that bilirubin binds to fibrinogen with moderate affinity. No red or blue shift of emission maximum of fibrinogen in the presence of bilirubin was observed, suggesting that the polarity around Trp and Tyr amino groups remains the same or almost the same. The proof that complex formation occurs and that fluorescence quenching is not a consequence of dynamic quenching came from UV-VIS spectrometry, Figure 5C. As it can be seen, especially in the presence of higher fibrinogen concentration, red shift of absorption maximum of bilirubin occurs and at the same time, the absorption spectra became wider.

In an aqueous solution, bilirubin exists in an equimolar ratio of M and P ridge tile conformers. Stereoselective binding of the P conformer of bilirubin to HSA induces optical activity or a positive Cotton effect in the CD spectra of the HSA-bilirubin complex [76]. This stereoselectivity does not exist in the case of bilirubin binding to fibrinogen. This means that fibrinogen's binding site(s) for bilirubin is not stereospecific and can bind both conformers.

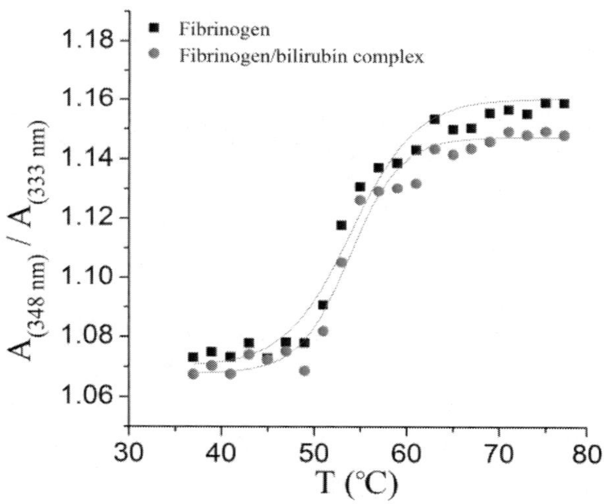

Figure 6. Determination of the melting point of fibrinogen and the fibrinogen/bilirubin complex; the ratio of fluorescence intensities at 348 and 333 nm (F348/F333) was calculated. The inflection point on the curve corresponds to the melting point. Reprinted from [75].Thermal denaturation of fibrinogen, similar to other proteins, causes red shift of its emission maxima and this property was used to analyze its melting point in the presence of bilirubin. The melting point of fibrinogen remained the same in the presence of bilirubin and was determined to be about 54°C, Figure 6. This indicates that the binding of bilirubin does not destabilize fibrinogen.

Change in the absorption spectra of bilirubin when bound to fibrinogen suggests that the conformation of bound bilirubin is changed. Spectrometric calculations predict that a transition of bilirubin from a ridge tile to more cyclic conformation is followed by a red shift of its absorption maximum [76]. This suggests that bilirubin is present in more cyclic conformation in the fibrinogen-bilirubin complex. On the other hand, conformational freedom of bilirubin is not reduced significantly upon binding to fibrinogen, as suggested by the lack of increase of fluorescence signal of bilirubin (results not shown).

The structure of fibrinogen, both its tertiary and secondary, remained unaltered in the presence of bound bilirubin, as determined by near- and far-UV circular dichroism (CD) spectrometry, Figure 7A and 7B. This result is in agreement with the melting point analysis.

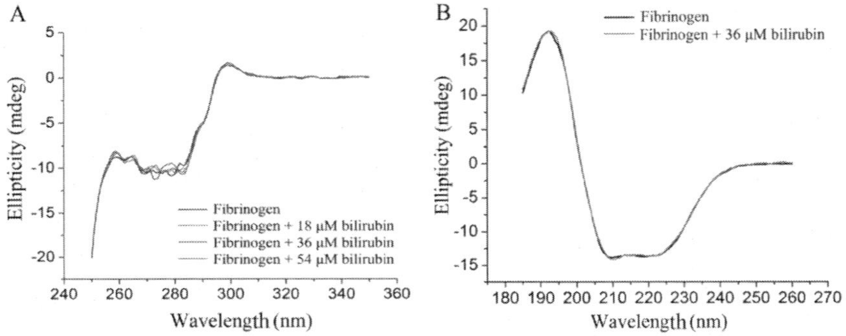

Figure 7. Near-UV (A) and far-UV (B) CD spectra of fibrinogen in the absence and the presence of bilirubin. Reprinted from [75].

Incubation with oxidative reagent, 2,2′-azobis(2-amidinopropane) dihydrochloride (AAPH), resulted in both oxidation of fibrinogen as well as the formation of protein aggregates, Figure 8A. Bilirubin was able to protect fibrinogen from aggregation, especially during the first hour of reaction and it was less effective in preventing aggregation after two hours, Figure 8A. Modification of fibrinogen carbonyls with dinitrophenylhydrazine (DNP) (products of oxidation) and subsequent detection using anti-DNP antibodies confirmed that the presence of bilirubin reduced their presence, suggesting that bilirubin protects fibrinogen from oxidation, Figures 8B and 8C. One characteristic of certain pathophysiological conditions is increased oxidative stress. Oxidation may significantly alter both the structure and function of proteins [77]. Fibrinogen is very susceptible to oxidation which usually gives it thrombogenic characteristics. For example, fibrinogen oxidation was observed in patients with myocardial infarction, cirrhosis or diabetes mellitus [78–80]. Oxidative stress affects the entire coagulation system causing hyperactivity of platelets as well as dysfunction of the endothelium [81]. The antioxidant effect of bilirubin was clearly demonstrated, as the presence of bilirubin protected fibrinogen from carbonylation and aggregation. Bilirubin is a potent antioxidant and it managed to protect even fibrinogen, a plasma protein most susceptible to oxidation [82]. Beneficial effects of mildly increased bilirubin like in Gilbert syndrome are already mentioned in the previous part; however, here it should be mentioned that protein oxidation, as well as concentration of fibrinogen, are both reduced with increased concentration of bilirubin [83, 84]. Bilirubin was also shown to be an endogenous antioxidant in human endothelial cells [85].

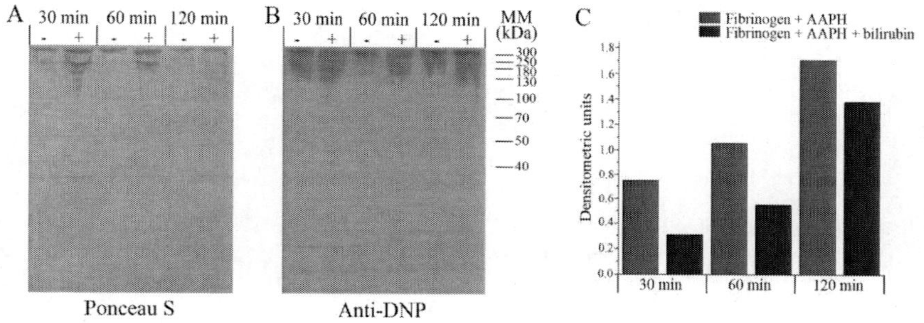

Figure 8. Determination of the susceptibility of fibrinogen to oxidation with AAPH in the absence (−) and in the presence (+) of bilirubin during 4 h. A native PAGE and a transfer to the PVDF membrane was performed, followed by the protein staining with Ponceau S (A), and derivatization of the protein carbonyls with DNP and immunoblotting with anti-DNP antibody (B). The ratio of densitometric signals obtained by immunoblotting and Ponceau S staining (C) (there was no protein detected by Ponceau S staining after 240 min, although an immunoreactive signal with anti-DNP antibody was recorded). Reprinted from [75].

Bilirubin had an effect on fibrin formation as well, Figure 9. In its presence, the initial speed of fibrin formation and coagulation time remained almost the same; however, maximal absorbance at the end of the fibrin formation process was higher in the presence of bilirubin, suggesting the formation of thicker fibrin fibers. Thicker fibrin fibers usually have higher permeability. However, other factors, such as oxidation, can influence its permeability [86, 87]. When permeability is reduced, plasmin cannot efficiently penetrate the fibrin clot, leading to a prolonged degradation time of fibrin and an increased risk of thrombus formation [77]. Oxidation reduces the permeability of fibrin due to the formation of hydrophobic patches that prevent the entry of plasmin from an aqueous environment [80].

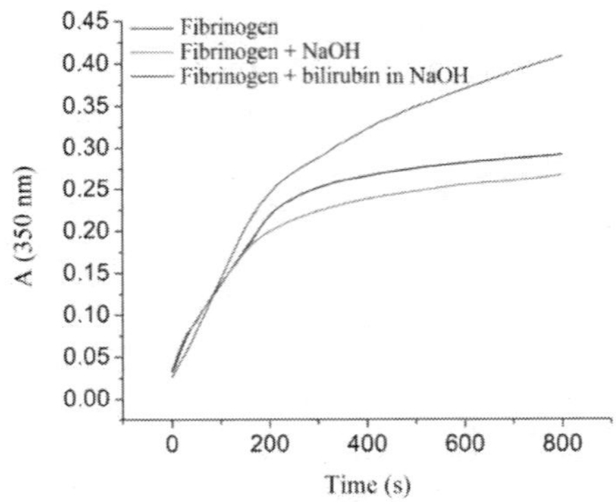

Figure 9. Determination of the effect of bilirubin on the formation of fibrin in the coagulation assay. Reprinted from [75].

Altogether, the results of this study show that bilirubin might be a beneficial modulator of hemostasis and an important protector of fibrinogen against oxidation [75]. It should be noted that the coagulation assay used in this work was based on the purified fibrinogen and commercially available thrombin. Many other important factors from blood influence fibrin formation. Nevertheless, fibrinogen and bilirubin are able to interact under physiological conditions. Bilirubin protects fibrinogen against oxidation and may modulate fibrin formation, which altogether suggests the possible physiological importance of this interaction.

Interactions of Bilirubin with Other proteins of Physiological Relevance

Interactions of Bilirubin with Ligandin

Glutathione S-Transferase or Ligandin is a cytosolic protein highly abundant in hepatocytes. Ligandin has a significant role in the hepatic uptake of bilirubin and its subsequent intracellular transport [88]. It is a homodimeric protein, while each subunit has two domains (α and β). Domain α binds glutathione, while hydrophobic substrates are accommodated in the pocket between two domains. However, the binding of hydrophobic non-substrate ligands, such as bilirubin, is distinct from the hydrophobic substrate-binding site [89]. The crystallographic analysis and fluorescence-resonance energy transfer approach proposes that the ordered water network binding site at the interface of subunits could accommodate hydrophobic non-substrate ligands [90]. However, there is still debate about the stoichiometry of bilirubin binding to ligandin, the binding affinity, and the influence of bilirubin binding on the catalytic activity of ligandin. Several studies demonstrated high affinity (~ 10^7 M^{-1}) binding of bilirubin to rat ligandin, and this high-affinity site is independent of the pocket at which catalytically reactive substrates bind [88, 91, 92]. Interestingly, the presence of glutathione could significantly promote the binding of bilirubin to rat ligandin [93]. Besides the high-affinity site, the presence of secondary bilirubin site on rat ligandin was observed with two orders of magnitude lower affinity. The same study proposes that bilirubin binding at the secondary site could competitively inhibit glutathione-S-transferase activity [91].

On the other hand, the characterization of bilirubin binding to human liver ligandin showed the absence of high-affinity binding, while the observed binding constant was in the range 10^5-10^6 M^{-1}. Furthermore, bilirubin binding to human ligandin is complex, comprising two interacting sites (cooperative binding) with a Hill coefficient of 1.5. Interestingly, bilirubin binding to human ligandin is not independent of glutathione S-transferase activity. Once bilirubin binds to ligandin, the initial complex is catalytically active. However, this formed complex undergoes conformational changes that induce catalytic activity loss. Interestingly, if bilirubin is added to ligandin in the presence of glutathione, the conformation change would also be induced, but obtained conformer will retain mostly catalytic activity [94].

Peroxisome Proliferator-Activated Receptor-A

Peroxisome proliferator-activated receptor (PPAR-α) is a ligand-activated transcriptional factor that belongs to the family of nuclear receptors [95]. It regulates the expression of genes involved in energy homeostasis, nutrient metabolism, immunity and inflammation [96]. Bilirubin binds with moderate affinity (~10^5 M^{-1}) at the well-known ligand-binding pocket of PPARα [97, 98]. Bilirubin binding enhances PPARα activity, which leads to the increase of lipid burning genes [98], which is in accordance with previous findings of the negative correlation between bilirubin levels in humans and obesity [99]. Therefore, bilirubin could be considered a metabolic hormone, making it a potential agent for addressing metabolic syndrome [97].

Cystatin

Cystatins are protein-based inhibitors of cysteine proteases of the papain family. They inactivate target proteases reversibly and competitively by indirect blockage of the catalytic centers of these enzymes, which prevents substrate binding and cleavage [100, 101]. Liver cystatin can bind bilirubin with moderate affinity (~10^5 M^{-1}). Bilirubin binding induces the transition of α-helix to β-sheet conformation followed by functional inactivation of cystatin, i.e., its inability to inhibit cysteine proteases. Inhibition of liver cystatin by bilirubin may father lead to increased cysteine proteases level, the hallmark of liver cirrhosis [100].

Prostaglandin D Synthase

Lipocalin-type prostaglandin D Synthase (PGDS) or β-Trace protein is a low-molecular-weight glycoprotein and one of the most important proteins of cerebrospinal fluid (CSF) [102]. Analysis of human CSF revealed the ability of bile pigments to interact with PGDS. Further experiments performed on recombinant PGDS confirmed high affinity (10^7-10^8 M^{-1}) binding of *P*-conformer of bilirubin in the hydrophobic pocket of protein [103, 104]. Additionally, strong binding of bilirubin induced the more compact structure of PGDS compared to free protein [105]. Therefore, PGDS may scavenge bilirubin under pathological conditions such as brain hemorrhages or blood-brain barriers, consequently inhibiting bilirubin-induced neuronal cell death [104, 105].

Collagen

Collagen is a fibrous structural protein and a major constituent of all extracellular matrices [107]. There are many types of collagens, but all collagens are characterized by the triple helix structure, formed by the association of three polypeptide chains [108]. Several studies demonstrated the ability of collagen to bind bilirubin [109-113]. The binding affinity of the collagen: bilirubin system is significantly lower (~10^3-10^4 M^{-1}) compared to high-affinity binding to HSA. However, considering that collagen is the most abundant protein in the human body, interactions with bilirubin could be of high physiological significance [109].

Denaturation of collagen by temperature or urea greatly diminish bilirubin binding, suggesting that conformation and organization of collagen fibrils strongly influence bilirubin binding [109]. Furthermore, Usharani and colleagues revealed specific bilirubin binding to $\alpha 1$(I) chain of type I collagen [110]. Based on the collagen-binding assays, the binding of bilirubin to collagen is found to be electrostatic, while bilirubin binding induces aggregation of collagen. Additionally, the collagen:bilirubin complex tends to undergo adsorption onto the surface of the fibroblast cells, which could have detrimental effects on fibroblasts proliferations [112]. Therefore, further characterization of collagen-bilirubin interaction is of high importance to design the drugs to inhibit the interface point of interaction between collagen and bilirubin [111, 112].

Interactions of Selected Bilirubin Precursors and Analogues with Proteins

Biliverdin

Biliverdin, Figure 1, a precursor of bilirubin, has a high propensity to interact with various proteins. Although biliverdin differs from bilirubin only by a double bond at the central carbon bridge, its presence gives biliverdin higher water solubility than bilirubin, which renders biliverdin nontoxic, while bilirubin is toxic to tissues. These subtle structural differences also cause the difference between the binding of bilirubin and biliverdin to HSA [56, 114]. Biliverdin binds to HSA with high affinity (10^6-10^7 M^{-1}). However, the same study using the same methodology (fluorescence quenching and enzymatic oxidation approaches) confirmed bilirubin binds to HSA with at least one order of magnitude higher affinity. Additionally, while bilirubin binds at more than one HSA site, the binding stoichiometry of HSA:biliverdin system is 1:1 [56]. Furthermore, CD spectroscopy revealed that biliverdin (M conformer) and bilirubin (P conformer) in the complex with HSA adopt different conformations [114, 115]. Meanwhile, similarly to the HSA:bilirubin system, the binding constant significantly depends on the presence of fatty acids: non-defatted albumin more strongly binds biliverdin than defatted HSA [56].

Like bilirubin, biliverdin binds with a high affinity to the lipocalin-type PGDS (10^7-10^8 M^{-1}), enabling PGDS to scavenge biliverdin in the CSF of patients with aneurysmal subarachnoid hemorrhage. On the other hand, biliverdin is not found in normal CSF [106]. Furthermore, biliverdin can interact with other lipocalin proteins such as α-Acid glycoprotein (AAG), an acute phase component of the human serum. Biliverdin:AAG system is characterized by moderate affinity binding (~10^5 M^{-1}) and stereoselective binding of biliverdin *P* conformer at the central hydrophobic pocket of AAG. However, increasing biliverdin hydrophobicity by esterifying its carboxyl groups enhances biliverdin affinity to AAG for two orders of magnitude. Hence, the esterified form of biliverdin could serve as a specific system for treating unwanted inflammatory responses mediated through PGDS action [116].

A recent study demonstrated the ability of biliverdin to interact with the N-terminal domain (NTD) of SARS-CoV-2 spike protein with high affinity (~10^8 M^{-1}). Biliverdin binding induces profound conformational changes in NTD significantly, leading to the suppression of the reactivity of SARS-CoV-2 spike with immune sera and inhibition of a subset of neutralizing antibodies. Therefore, it would be important to evaluate spike constructs deficient for the interaction with biliverdin as vaccine candidates [117].

Phycocyanobilin

Phycocyanobilin (PCB) is an open-chain tetrapyrrole chromophore (Figure 10A) with high structural similarity with biliverdin and bilirubin [54]. The major source of PCB is cyanobacteria Spirulina, a well-known food supplement with numerous health benefits [118, 119]. It has been documented that powerful health-promoting activities of Spirulina arise from the presence of the phycobiliprotein C-phycocyanin (C-PC), which covalently binds PCB [118]. A large number of studies have shown that C-PC exhibits significant anti-inflammatory, immunomodulatory, anti-cancer, nephroprotective and hepatoprotective effects [120], ascribed to antioxidant and radical scavenging properties of PCB [118]. Indeed, recent studies have shown that PCB exhibits significant anti-cancer [121], anti-inflammatory [122], atheroprotective [123], neuroprotective [124] and nephroprotective effects [125].

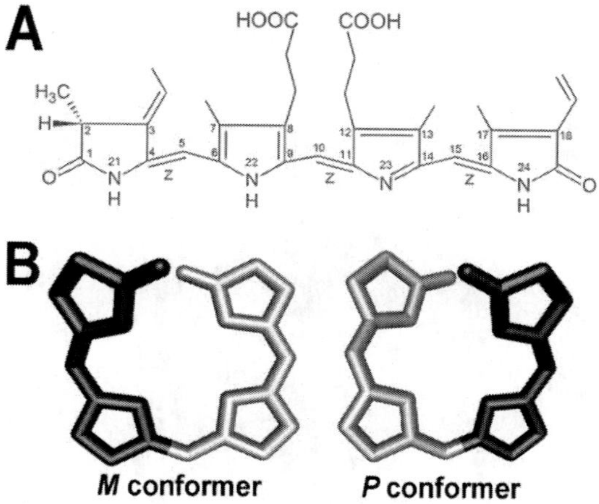

Figure. 10. (A) Chemical structure of PCB; (B) Schematic representation of opposite helical conformations (*M* or *P*) of PCB: half of the molecules drawn by bold are above the plane of the paper. Ring substituents were omitted for clarity; Reprinted from [54].

Considering the strong bioactive effects of PCB and its structural similarity with bilirubin, in our previous study, we characterised the binding of PCB to HSA. Results obtained by fluorescence and circular dichroism measurements confirmed high affinity (binding constant of 2.2 x 10^6 M^{-1}) and stereoselective binding of PCB *M*-conformer (Figure 10B) to HSA. Using the same experimental approach, the same study revealed 2-3 times stronger binding of bilirubin to HSA. A molecular docking study demonstrated two putative high-affinity binding pockets (subdomain IB and IIA) on HSA of virtually identical binding scores (Figure 11). Proposed binding sites are identical to the two previously identified binding sites for bilirubin. The location of the PCB binding site was additionally conformed by fluorescence spectroscopy in competition experiments with site-marker compounds such as hemin (subdomain IB) and warfarin (subdomain IIA). Moreover, the same experimental approach demonstrated the ability of PCB to displace bilirubin from the HSA:bilirubin complex and *vice versa* (Figure 12). Hence, in conditions characterized by an increased bilirubin plasma concentration or intake of drugs binding to IB or IIA binding site, the pharmacokinetics of PCB may also be changed [54].

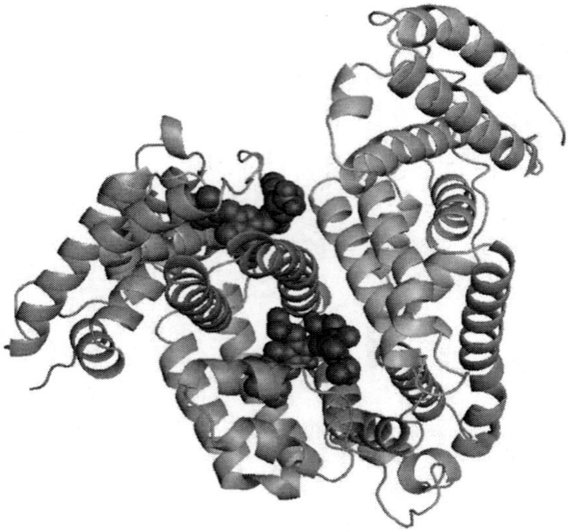

Figure 11. Molecular modeling of PCB binding to HSA. The domains in ribbon model of protein HSA (PDB ID: 1BM0) are color codes as follows: I, orange; II, green; III, turquoise blue. Structural models depicting two potential binding sites for bilirubin to HSA: bound bilirubin is shown in purple color space filling representation. Reprinted from [54].

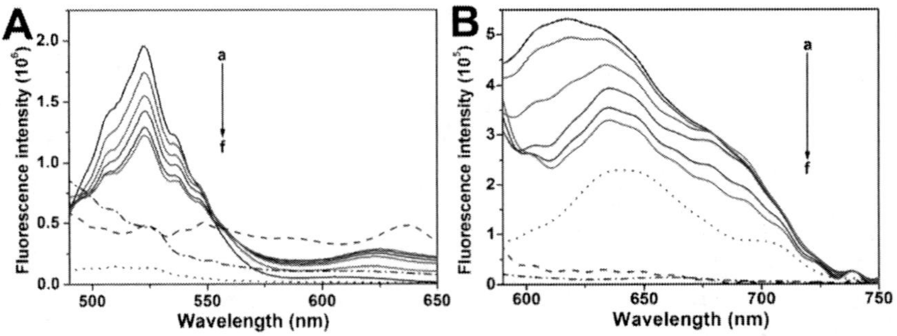

Figure 12. (A) Quenching of HSA–bilirubin complex (4 μM both) by PCB (0, 2, 4, 6, 8 and 10 μM, for curves a to f, respectively) using synchronous fluorescence spectroscopy, Δλ 56 nm. Dash line curve: 4 μM HSA in presence of 4 μM phycocyanobilin. Dot line curve: 10 μM bilirubin. Dash-dot line curve: 10 μM phycocyanobilin; (B) quenching of HSA–PCB complex (2 μM both) by bilirubin (0, 1, 2, 3, 4 and 5 μM, for curves a to f, respectively) using synchronous fluorescence spectroscopy, Δλ 87 nm. Dash line curve represents 2 μM HSA in the presence of 2 μM bilirubin. Dot line curve: 2 μM PCB. Dash-dot line curve represents 10 μM bilirubin. Reprinted from [54].

Furthermore, the binding of PCB to HSA induces structural changes in the protein and the ligand. UV-visible absorbance measurements indicate that bound PCB has more stretched conformation than free form. Experimental data have shown an increase in thermal and proteolytic stability, as well as the α-helical content of the HSA upon PCB binding. Therefore, the binding of PCB to HSA may provide higher protein stability and longevity in plasma [126, 127].

Our recent study, combining multi-spectroscopic and molecular docking approaches, characterized the binding of PCB to bovine catalase, a life-essential antioxidant enzyme. Considering the high sensitivity of PCB to oxidation, its binding to catalase could reduce free-radical induced chromophore oxidation. Fluorescence quenching experiments demonstrated moderate binding (Ka of 3.9×10^4 M^{-1}) of PCB to catalase. The presence of tetrapyrrole chromophore does not significantly influence secondary structure content or the protein's thermal stability. However, changes in the tertiary structure of catalase were detected upon ligand binding but without adverse effects on enzyme activity. The molecular docking study proposed the binding of PCB near the cavity between the enzyme homotetramer subunits (Figure 13). Finally, complex formation protects the pigment from free-radical induced oxidation suggesting possible prolonging its half-life and bioactivity *in vivo* in the complex with catalase [128].

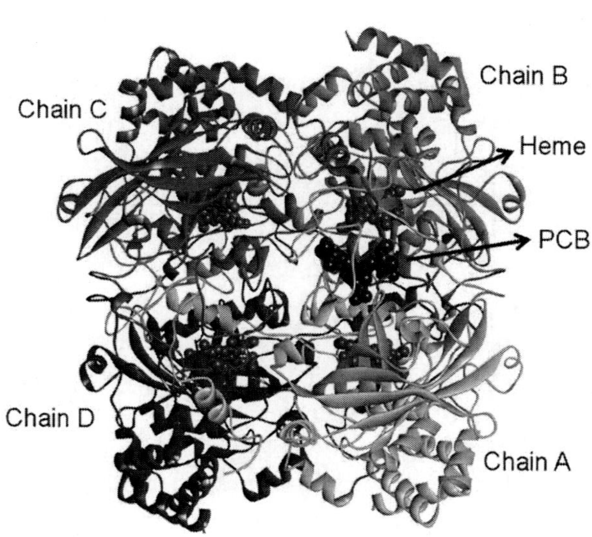

Figure 13. Docking results of the catalase (CAT)/phycocyanobilin (PCB) complex. Reprinted from [128].

COVID-19 pandemic has directed one part of research on the antiviral activities of natural, food-derived, bioactive compounds. *In silico* approach is mainly used to detect the possible interactions between bioactive ligands and SARS-CoV-2 proteins [129]. Recent studies employed molecular docking study to demonstrate the ability of PCB to interact with the key enzymes of SARS-CoV-2: the Main protease (M^{pro}) and RNA-dependent RNA polymerase, which suggests the PCB's ability to inhibit these enzymes [130, 131]. Indeed, *in vitro* study conducted on M^{pro} and papain-like protease (PL^{pro}) of SARS-CoV-2 confirmed the inhibitory potential of PCB with the corresponding IC_{50} values of 71 and 62 μM, respectively [132]. Besides interactions with the key viral enzymes, PCB can interact with the receptor-binding domain (RBD) of SARS-CoV-2 spike protein [133]. RBD binds human angiotensin-converting enzyme 2 (ACE2), enabling virus attachment to the host cell [134]. Interestingly, PCB interacts with the RBD amino acid residues involved in the binding to ACE2, suggesting PCB's potential to inhibit interactions between RBD and ACE2, and consequently infection of human cells by SARS-CoV-2 virus [133].

Conclusion

Presented work gave an overview of bilirubin metabolism and its binding characteristics for several proteins. Both proteins from circulation and those in tissues play an important role in bilirubin metabolism. While unconjugated bilirubin is predominantly bound to HSA in normal conditions, different pathologic conditions may influence the circulatory concentrations of both binding partners. Under those conditions, interactions of bilirubin with other proteins, like fibrinogen or alpha-2-macrogolbulin for example, may be more significant. Free unconjugated bilirubin is found in very small amount in circulation and while marginal increases of this molecule are considered to be beneficial, significant increases may cause serious conditions, including death. If HSA is saturated by bilirubin, other proteins may bind it and thus possibly mitigate toxic effects of free unconjugated bilirubin. Further studies are needed to explore this more deeply. Analogues of bilirubin are also mentioned and some of them like phycocyanobilin are found in food sources. Interactions of PCB with HSA and catalase are described here, with potential protective anti-oxidative effect of PCB toward these proteins. Further studies are needed to describe pharmacokinetic and pharmacodynamic properties of PCB in organism. Understanding characteristics of its binding for both circulatory and tissue proteins is very important for this.

References

[1] Wang, X., J.R. Chowdhury, N.R. Chowdhury, Bilirubin metabolism: Applied physiology, *Current Paediatrics.* 16 (2006) 70–74. https://doi.org/10.1016/j.cupe.2005.10.002.

[2] Billing, B.H. Bilirubin Metabolism and Jaundice with Special Reference to Unconjugated Hyperbilirubinaemia, Annals of Clinical Biochemistry: *International Journal of Laboratory Medicine.* 7 (1970) 69–74. https://doi.org/10.1177/000456327000700401.

[3] Hansen, T.W.R. Core Concepts: Bilirubin Metabolism, *NeoReviews.* 11 (2010) e316–e322. https://doi.org/10.1542/neo.11-6-e316.

[4] Ritter, J.K., F. Chen, Y.Y. Sheen, H.M. Tran, S. Kimura, M.T. Yeatman, I.S. Owens, A novel complex locus UGT1 encodes human bilirubin, phenol, and other UDP-glucuronosyltransferase isozymes with identical carboxyl termini., *Journal of Biological Chemistry.* 267 (1992) 3257–3261. https://doi.org/10.1016/S0021-9258(19)50724-4.

[5] Borst, P., R.O. Elferink, Mammalian ABC Transporters in Health and Disease, *Annual Review of Biochemistry.* 71 (2002) 537–592. https://doi.org/10.1146/annurev.biochem.71.102301.093055.

[6] Guerra Ruiz, A.R., J. Crespo, R.M. López Martínez, P. Iruzubieta, G. Casals Mercadal, M. Lalana Garcés, B. Lavin, M. Morales Ruiz, Measurement and clinical usefulness of bilirubin in liver disease, Advances in Laboratory Medicine / *Avances En Medicina de Laboratorio.* 2 (2021) 352–361. https://doi.org/10.1515/almed-2021-0047.

[7] Pellock, S.J., M.R. Redinbo, Glucuronides in the gut: Sugar-driven symbioses between microbe and host, *Journal of Biological Chemistry.* 292 (2017) 8569–8576. https://doi.org/10.1074/jbc.R116.767434.

[8] Sticova, E., A. Lodererova, E. van de Steeg, S. Frankova, M. Kollar, V. Lanska, R. Kotalova, T. Dedic, A.H. Schinkel, M. Jirsa, Down-regulation of OATP1B proteins correlates with hyperbilirubinemia in advanced cholestasis., International *Journal of Clinical and Experimental Pathology.* 8 (2015) 5252–5262.

[9] Erlinger, S., I.M. Arias, D. Dhumeaux, Inherited Disorders of Bilirubin Transport and Conjugation: New Insights into Molecular Mechanisms and Consequences, *Gastroenterology.* 146 (2014) 1625–1638. https://doi.org/10.1053/j.gastro.2014.03.047.

[10] van de Steeg, E., V. Stránecký, H. Hartmannová, L. Nosková, M. Hřebíček, E. Wagenaar, A. van Esch, D.R. de Waart, R.P.J. Oude Elferink, K.E. Kenworthy, E. Sticová, M. al-Edreesi, A.S. Knisely, S. Kmoch, M. Jirsa, A.H. Schinkel, Complete OATP1B1 and OATP1B3 deficiency causes human Rotor syndrome by interrupting

conjugated bilirubin reuptake into the liver, *Journal of Clinical Investigation.* 122 (2012) 519–528. https://doi.org/10.1172/JCI59526.

[11] Treviño, L.R., N. Shimasaki, W. Yang, J.C. Panetta, C. Cheng, D. Pei, D. Chan, A. Sparreboom, K.M. Giacomini, C.-H. Pui, W.E. Evans, M. v. Relling, Germline Genetic Variation in an Organic Anion Transporter Polypeptide Associated With Methotrexate Pharmacokinetics and Clinical Effects, *Journal of Clinical Oncology.* 27 (2009) 5972–5978. https://doi.org/10.1200/JCO.2008.20.4156.

[12] Morimoto, K., T. Oishi, S. Ueda, M. Ueda, M. Hosokawa, K. Chiba, A Novel Variant Allele of OATP-C (SLCO1B1) Found in a Japanese Patient with Pravastatin-induced Myopathy, *Drug Metabolism and Pharmacokinetics.* 19 (2004) 453–455. https://doi.org/10.2133/dmpk.19.453.

[13] Rowland, A., J.O. Miners, P.I. Mackenzie, The UDP-glucuronosyltransferases: Their role in drug metabolism and detoxification, *The International Journal of Biochemistry & Cell Biology.* 45 (2013) 1121–1132. https://doi.org/10.1016/j.biocel.2013.02.019.

[14] Bosma, P.J., J.R. Chowdhury, C. Bakker, S. Gantla, A. de Boer, B.A. Oostra, D. Lindhout, G.N.J. Tytgat, P.L.M. Jansen, R.P.J.O. Elferink, N.R. Chowdhury, The Genetic Basis of the Reduced Expression of Bilirubin UDP-Glucuronosyltransferase 1 in Gilbert's Syndrome, *New England Journal of Medicine.* 333 (1995) 1171–1175. https://doi.org/10.1056/NEJM199511023331802.

[15] Matsui, K., Y. Maruo, H. Sato, Y. Takeuchi, Combined effect of regulatory polymorphisms on transcription of UGT1A1 as a cause of Gilbert syndrome, *BMC Gastroenterology.* 10 (2010) 57. https://doi.org/10.1186/1471-230X-10-57.

[16] Kaplan, M., C. Hammerman, F.F. Rubaltelli, M.T. Vilei, E. Levy-Lahad, P. Renbaum, H.J. Vreman, D.K. Stevenson, M. Muraca, Hemolysis and bilirubin conjugation in association with UDP-glucuronosyltransferase 1A1 promoter polymorphism, *Hepatology.* 35 (2002) 905–911. https://doi.org/10.1053/jhep.2002.32526.

[17] K.M. Korenblat, P.D. Berk, Hyperbilirubinemia in the setting of antiviral therapy, *Clinical Gastroenterology and Hepatology.* 3 (2005) 303–310. https://doi.org/10.1016/S1542-3565(05)00083-2.

[18] C.P. Strassburg, Hyperbilirubinemia syndromes (Gilbert-Meulengracht, Crigler-Najjar, Dubin-Johnson, and Rotor syndrome), *Best Practice & Research Clinical Gastroenterology.* 24 (2010) 555–571. https://doi.org/10.1016/j.bpg.2010.07.007.

[19] M. Rotger, P. Taffé, G. Bleiber, H.F. Günthard, H. Furrer, P. Vernazza, H. Drechsler, E. Bernasconi, M. Rickenbach, A. Telenti, Gilbert Syndrome and the Development of Antiretroviral Therapy–Associated Hyperbilirubinemia, *The Journal of Infectious Diseases.* 192 (2005) 1381–1386. https://doi.org/10.1086/466531.

[20] Liu, X., D. Cheng, Q. Kuang, G. Liu, W. Xu, Association of UGT1A1*28 polymorphisms with irinotecan-induced toxicities in colorectal cancer: a meta-analysis in Caucasians, *The Pharmacogenomics Journal.* 14 (2014) 120–129. https://doi.org/10.1038/tpj.2013.10.

[21] Biason, P., S. Masier, G. Toffoli, UGT1A1*28 and Other UGT1A Polymorphisms as Determinants of Irinotecan Toxicity, *Journal of Chemotherapy.* 20 (2008) 158–165. https://doi.org/10.1179/joc.2008.20.2.158.

[22] Watchko, J.F., C. Tiribelli, Bilirubin-Induced Neurologic Damage — Mechanisms and Management Approaches, *New England Journal of Medicine.* 369 (2013) 2021–2030. https://doi.org/10.1056/NEJMra1308124.

[23] Canu, G., A. Minucci, C. Zuppi, E. Capoluongo, Gilbert and Crigler Najjar syndromes: An update of the UDP-glucuronosyltransferase 1A1 (UGT1A1) gene mutation database, *Blood Cells, Molecules, and Diseases.* 50 (2013) 273–280. https://doi.org/10.1016/j.bcmd.2013.01.003.

[24] van der Veere, C.N., M. Sinaasappel, A.F. McDonagh, P. Rosenthal, P. Labrune, M. Odiévre, J. Fevery, J. Otte, P. McClean, G. Bürk, V. Masakowski, W. Sperl, A.P. Mowat, G.M. Vergani, K. Heller, J.P. Wilson, R. Shepherd, P.L. Jansen, Current therapy for Crigler-Najjar syndrome type 1: Report of a world registry, *Hepatology.* 24 (1996) 311–315. https://doi.org/10.1002/hep.510240205.

[25] Bortolussi, G., L. Zentilin, G. Baj, P. Giraudi, C. Bellarosa, M. Giacca, C. Tiribelli, A.F. Muro, Rescue of bilirubin-induced neonatal lethality in a mouse model of Crigler-Najjar syndrome type I by AAV9-mediated gene transfer, *The FASEB Journal.* 26 (2012) 1052–1063. https://doi.org/10.1096/fj.11-195461.

[26] Gordon, E.R., E.A. Shaffer, A. Sass-Kortsak, Bilirubin Secretion and Conjugation in the Crigler-Najjar Syndrome Type II, *Gastroenterology.* 70 (1976) 761–765. https://doi.org/10.1016/S0016-5085(76)80271-5.

[27] Dubin, I.N., F.B. Johnson, Chronic idiopathic jaundice with unidentified pigment in liver cells, *Medicine* (Baltimore). 33 (1954) 155–197. https://doi.org/10.1097/00005792-195409000-00001.

[28] Frank, M., M. Doss, D.G. de Carvalho, Diagnostic and pathogenetic implications of urinary coproporphyrin excretion in the Dubin-Johnson syndrome, *Hepato-Gastroenterology. 37* (1990) 147–151.

[29] Shani, M., U. Seligsohn, E. Gilon, C. Sheba, A. Adam, Dubin-Johnson syndrome in Israel. I. Clinical, laboratory, and genetic aspects of 101 cases, *The Quarterly Journal of Medicine.* 39 (1970) 549–567.

[30] Togawa, T., T. Mizuochi, T. Sugiura, H. Kusano, K. Tanikawa, T. Sasaki, F. Ichinose, S. Kagimoto, T. Tainaka, H. Uchida, S. Saitoh, Clinical, Pathologic, and Genetic Features of Neonatal Dubin-Johnson Syndrome: A Multicenter Study in Japan, *The Journal of Pediatrics.* 196 (2018) 161–167. https://doi.org/10.1016/j.jpeds.2017.12.058.

[31] Gazzin, S., F. Masutti, L. Vitek, C. Tiribelli, The molecular basis of jaundice: An old symptom revisited, Liver International. 37 (2017) 1094–1102. https://doi.org/10.1111/liv.13351.

[32] Wakusawa, S., I. Machida, S. Suzuki, H. Hayashi, M. Yano, K. Yoshioka, Identification of a novel 2026G→C mutation of the MRP2 gene in a Japanese patient with Dubin-Johnson syndrome, *Journal of Human Genetics.* 48 (2003) 425–429. https://doi.org/10.1007/s10038-003-0052-0.

[33] Wu, L., W. Zhang, S. Jia, X. Zhao, D. Zhou, A. Xu, W. Duan, Z. Wu, H. Li, S. Zheng, Y. Nan, J. Jia, J. Huang, X. Ou, Mutation analysis of the ABCC2 gene in

Chinese patients with Dubin-Johnson syndrome, *Experimental and Therapeutic Medicine. 16* (2018) 4201–4206. https://doi.org/10.3892/etm.2018.6682.

[34] Hashimoto, K., T. Uchiumi, T. Konno, T. Ebihara, T. Nakamura, M. Wada, S. Sakisaka, F. Maniwa, T. Amachi, K. Ueda, M. Kuwano, Trafficking and functional defects by mutations of the ATP-binding domains in MRP2 in patients with Dubin-Johnson syndrome, *Hepatology.* 36 (2002) 1236–1245. https://doi.org/10.1053/jhep.2002.36368.

[35] Keitel, V., A.T. Nies, M. Brom, J. Hummel-Eisenbeiss, H. Spring, D. Keppler, A common Dubin-Johnson syndrome mutation impairs protein maturation and transport activity of MRP2 (ABCC2), *American Journal of Physiology-Gastrointestinal and Liver Physiology.* 284 (2003) G165–G174. https://doi.org/10.1152/ajpgi.00362.2002.

[36] Boon, A.C., C.L. Hawkins, J.S. Coombes, K.H. Wagner, A.C. Bulmer, Bilirubin scavenges chloramines and inhibits myeloperoxidase-induced protein/lipid oxidation in physiologically relevant hyperbilirubinemic serum, *Free Radical Biology and Medicine.* 86 (2015) 259–268. https://doi.org/10.1016/j.freeradbiomed.2015.05.031.

[37] Wagner, K.-H., M. Wallner, C. Mölzer, S. Gazzin, A.C. Bulmer, C. Tiribelli, L. Vitek, Looking to the horizon: the role of bilirubin in the development and prevention of age-related chronic diseases, *Clinical Science.* 129 (2015) 1–25. https://doi.org/10.1042/CS20140566.

[38] Gazzin, S., L. Vitek, J. Watchko, S.M. Shapiro, C. Tiribelli, A Novel Perspective on the Biology of Bilirubin in Health and Disease, *Trends in Molecular Medicine.* 22 (2016) 758–768. https://doi.org/10.1016/j.molmed.2016.07.004.

[39] Kundur, A.R., I. Singh, A.C. Bulmer, Bilirubin, platelet activation and heart disease: A missing link to cardiovascular protection in Gilbert's syndrome?, *Atherosclerosis.* 239 (2015) 73–84. https://doi.org/10.1016/j.atherosclerosis.2014.12.042.

[40] Hinds, T.D., S.O. Adeosun, A.A. Alamodi, D.E. Stec, Does bilirubin prevent hepatic steatosis through activation of the PPARα nuclear receptor?, *Medical Hypotheses.* 95 (2016) 54–57. https://doi.org/10.1016/j.mehy.2016.08.013.

[41] Stec, D.E., K. John, C.J. Trabbic, A. Luniwal, M.W. Hankins, J. Baum, T.D. Hinds, Bilirubin Binding to PPARα Inhibits Lipid Accumulation, *PLOS ONE.* 11 (2016) e0153427. https://doi.org/10.1371/journal.pone.0153427.

[42] Basiglio, C.L., S.M. Arriaga, F. Pelusa, A.M. Almará, J. Kapitulnik, A.D. Mottino, Complement activation and disease: protective effects of hyperbilirubinaemia, *Clinical Science.* 118 (2010) 99–113. https://doi.org/ 10.1042/CS20080540.

[43] Mazzone, G.L., I. Rigato, J.D. Ostrow, F. Bossi, A. Bortoluzzi, C.H.C. Sukowati, F. Tedesco, C. Tiribelli, Bilirubin inhibits the TNFα-related induction of three endothelial adhesion molecules, *Biochemical and Biophysical Research Communications.* 386 (2009) 338–344. https://doi.org/10.1016/j.bbrc.2009.06.029.

[44] Mazzone, G.L., I. Rigato, J.D. Ostrow, C. Tiribelli, Bilirubin effect on endothelial adhesion molecules expression is mediated by the NF-kappaB signaling pathway., *Bioscience Trends.* 3 (2009) 151–157.

[45] Fujii, M., T. Inoguchi, S. Sasaki, Y. Maeda, J. Zheng, K. Kobayashi, R. Takayanagi, Bilirubin and biliverdin protect rodents against diabetic nephropathy by downregulating NAD(P)H oxidase, *Kidney International.* 78 (2010) 905–919. https://doi.org/10.1038/ki.2010.265.

[46] Gazzin, S., F. Masutti, L. Vitek, C. Tiribelli, The molecular basis of jaundice: An old symptom revisited, *Liver International.* 37 (2017) 1094–1102. https://doi.org/10.1111/liv.13351.

[47] Stec, D.E., P.A. Hosick, J.P. Granger, Bilirubin, Renal Hemodynamics, and Blood Pressure, *Frontiers in Pharmacology.* 3 (2012) 18. https://doi.org/10.3389/fphar.2012.00018.

[48] Pflueger, A., A.J. Croatt, T.E. Peterson, L.A. Smith, L. v. d'Uscio, Z.S. Katusic, K.A. Nath, The hyperbilirubinemic Gunn rat is resistant to the pressor effects of angiotensin II, *American Journal of Physiology-Renal Physiology.* 288 (2005) F552–F558. https://doi.org/10.1152/ajprenal.00278.2004.

[49] Takeda, T., A. Mu, T.T. Tai, S. Kitajima, S. Taketani, Continuous de novo biosynthesis of haem and its rapid turnover to bilirubin are necessary for cytoprotection against cell damage, *Scientific Reports.* 5 (2015) 10488. https://doi.org/10.1038/srep10488.

[50] Zucker, S.D., P.S. Horn, K.E. Sherman, Serum bilirubin levels in the U.S. population: Gender effect and inverse correlation with colorectal cancer, *Hepatology.* 40 (2004) 827–835. https://doi.org/10.1002/hep.20407.

[51] Ollinger, R., P. Kogler, J. Troppmair, M. Hermann, M. Wurm, A. Drasche, I. Konigsrainer, A. Amberger, H. Weiss, D. Ofner, F.H. Bach, R. Margreiter, Bilirubin Inhibits Tumor Cell Growth via Activation of ERK, *Cell Cycle.* 6 (2007) 3078–3085. https://doi.org/10.4161/cc.6.24.5022.

[52] Varshney, A., P. Sen, E. Ahmad, M. Rehan, N. Subbarao, R.H. Khan, Ligand binding strategies of human serum albumin: how can the cargo be utilised?, Chirality 22(1) (2010) 77-87.https://doi.org: 10.1002/chir.20709.

[53] Petersen, C.E., C.E. Ha, K. Harohalli, J.B. Feix, N.V. Bhagavan, A dynamic model for bilirubin binding to human serum albumin, *J Biol Chem* 275(28) (2000) 20985-95.https://doi.org: 10.1074/jbc.M001038200.

[54] Minic, S.L., M. Milcic, D. Stanic-Vucinic, M. Radibratovic, T.G. Sotiroudis, M.R. Nikolic, T.Ć. Velickovic, Phycocyanobilin, a bioactive tetrapyrrolic compound of blue-green alga Spirulina, binds with high affinity and competes with bilirubin for binding on human serum albumin, *RSC Advances* 5(76) (2015) 61787-61798.https://doi.org: 10.1039/c5ra05534b.

[55] Minomo, A., Y. Ishima, U. Kragh-Hansen, V.T. Chuang, M. Uchida, K. Taguchi, H. Watanabe, T. Maruyama, H. Morioka, M. Otagiri, Biological characteristics of two lysines on human serum albumin in the high-affinity binding of 4Z,15Z-bilirubin-IXalpha revealed by phage display, *FEBS J* 278(21) (2011) 4100-11.https://doi.org: 10.1111/j.1742-4658.2011.08316.x.

[56] Ahlfors, C. Competitive Interaction of Biliverdin and Bilirubin Only at the Primary Bilirubin Binding Site on Human Albumin, *Analytical Biochemistry* 110 (1981) 295-307.https://doi.org/10.1016/0003-2697(81)90195-0.

[57] Zunszain, P.A., J. Ghuman, A.F. McDonagh, S. Curry, Crystallographic analysis of human serum albumin complexed with 4Z,15E-bilirubin-IXalpha, *J Mol Biol* 381(2) (2008) 394-406.https://doi.org: 10.1016/j.jmb.2008.06.016.

[58] Goncharova, I., S. Orlov, M. Urbanova, The location of the high- and low-affinity bilirubin-binding sites on serum albumin: ligand-competition analysis investigated by circular dichroism, *Biophys Chem* 180-181 (2013) 55-65.https://doi.org: 10.1016/j.bpc.2013.06.004.

[59] Ascenzi, P., A. Bocedi, S. Notari, E. Menegatti, M. Fasano, Heme impairs allosterically drug binding to human serum albumin Sudlow's site I, *Biochem Biophys Res Commun* 334(2) (2005) 481-6.https://doi.org: 10.1016/j.bbrc. 2005.06.127.

[60] Zsila, F., Subdomain IB is the third major drug binding region of human serum albumin: toward the three-sites model, *Mol Pharm* 10(5) (2013) 1668-82.https://doi.org: 10.1021/mp400027q.

[61] Morioka, I., S. Iwatani, T. Koda, K. Iijima, H. Nakamura, Disorders of bilirubin binding to albumin and bilirubin-induced neurologic dysfunction, *Semin Fetal Neonatal Med* 20(1) (2015) 31-36.https://doi.org: 10.1016/j.siny.2014.11.001.

[62] Odell, G.B., J.O. Cukier, E.M. Ostrea, Jr., A.C. Maglalang, R.L. Poland, The influence of fatty acids on the binding of bilirubin to albumin, *J Lab Clin Med* 89(2) (1977) 295-307.https://doi.org/10.5555/uri:pii:0022214377900890.

[63] Amoorahim, M., M.R. Ashrafi-Kooshk, S. Esmaeili, M. Shahlaei, S. Moradi, R. Khodarahmi, Physiological changes in the albumin-bound non-esterified free fatty acids critically influence heme/bilirubin binding properties of the protein: A comparative, in vitro, spectroscopic study using the endogenous biomolecules, *Spectrochim Acta A Mol Biomol Spectrosc* 235 (2020) 118298.https://doi.org: 10.1016/j.saa.2020.118298.

[64] Maruyama, K., S. Harada, H. Nishigori, M. Iwatsuru, Classification of drugs on the basis of bilirubin-displacing effect on human serum albumin, *Chem Pharm Bull* (Tokyo) 32(6) (1984) 2414-20.https://doi.org: 10.1248/cpb.32.2414.

[65] Terentiev, A.A., N.T. Moldogazieva, Alpha-fetoprotein: a renaissance, *Tumour Biol* 34(4) (2013) 2075-91.https://doi.org: 10.1007/s13277-013-0904-y.

[66] Aoyagi, Y., T. Ikenaka, F. Ichida, alpha-Fetoprotein as a carrier protein in plasma and its bilirubin-binding ability, *Cancer Res* 39(9) (1979) 3571-4.https://.

[67] Berde, C.B., M. Nagai, H.F. Deutsch, Human alpha-fetoprotein. Fluorescence studies on binding and proximity relationships for fatty acids and bilirubin, *J Biol Chem* 254(24) (1979) 12609-14.https://.

[68] Rassart, E., A. Bedirian, S. Do Carmo, O. Guinard, J. Sirois, L. Terrisse, R. Milne, Apolipoprotein D, *Biochim Biophys Acta* 1482(1-2) (2000) 185-98.https://doi.org: 10.1016/s0167-4838(00)00162-x.

[69] Eichinger, A., A. Nasreen, H.J. Kim, A. Skerra, Structural insight into the dual ligand specificity and mode of high density lipoprotein association of apolipoprotein D, *J Biol Chem* 282(42) (2007) 31068-75.https://doi.org: 10.1074/ jbc.M703552200.

[70] Peitsch, M.C., M.S. Boguski, Is apolipoprotein D a mammalian bilin-binding protein?, *New Biol* 2(2) (1990) 197-206.https://.

[71] Goessling, W., S.D. Zucker, Role of apolipoprotein D in the transport of bilirubin in plasma, *Am J Physiol Gastrointest Liver Physiol* 279(2) (2000) G356-65.https://doi.org: 10.1152/ajpgi.2000.279.2.G356.

[72] Cooke, J.R., L.B. Roberts, The binding of bilirubin to serum proteins, *Clin Chim Acta* 26(3) (1969) 425-36.https://doi.org: 10.1016/0009-8981(69)90070-9.

[73] Sunderic, M., T. Vasovic, M. Milcic, C. Miljevic, O. Nedic, M.R. Nikolic, N. Gligorijevic, Antipsychotic clozapine binding to alpha-2-macroglobulin protects interacting partners against oxidation and preserves the anti-proteinase activity of the protein, *Int J Biol Macromol* 183 (2021) 502-512.https://doi.org: 10.1016/j.ijbiomac.2021.04.155.

[74] Ali, S.S., M.K. Zia, T. Siddiqui, H. Ahsan, F.H. Khan, Bilirubin binding affects the structure and function of alpha-2-macroglobulin, *J Immunoassay Immunochem* 41(5) (2020) 841-851.https://doi.org: 10.1080/15321819.2020.1783290.

[75] Gligorijević, N., S. Minić, D. Robajac, M. Nikolić, T. Ćirković Veličković, O. Nedić, Characterisation and the effects of bilirubin binding to human fibrinogen, *Int. J. Biol. Macromol.* 128 (2019) 74–79. https://doi.org/10.1016/j.ijbiomac.2019.01.124.

[76] Petersen, C.E., C.-E. Ha, K. Harohalli, J.B. Feix, N.V. Bhagavan, A Dynamic Model for Bilirubin Binding to Human Serum Albumin, J. *Biol. Chem.* 275 (2000) 20985–20995. https://doi.org/10.1074/jbc.M001038200.

[77] Curtis, J.M., W.S. Hahn, E.K. Long, J.S. Burrill, E.A. Arriaga, D.A. Bernlohr, Protein carbonylation and metabolic control systems, *Trends Endocrinol. Metab.* 23 (2012) 399–406. https://doi.org/10.1016/j.tem.2012.05.008.

[78] Becatti, M., R. Marcucci, G. Bruschi, N. Taddei, D. Bani, A.M. Gori, B. Giusti, G.F. Gensini, R. Abbate, C. Fiorillo, Oxidative Modification of Fibrinogen Is Associated With Altered Function and Structure in the Subacute Phase of Myocardial Infarction, Arterioscler. *Thromb. Vasc. Biol.* 34 (2014) 1355–1361. https://doi.org/10.1161/ATVBAHA.114.303785.

[79] Dunn, E.J., H. Philippou, R.A.S. Ariëns, P.J. Grant, Molecular mechanisms involved in the resistance of fibrin to clot lysis by plasmin in subjects with type 2 diabetes mellitus, *Diabetologia.* 49 (2006) 1071–1080. https://doi.org/10.1007/s00125-006-0197-4.

[80] Lisman, T., R.A.S. Ariëns, Alterations in Fibrin Structure in Patients with Liver Diseases, Semin. *Thromb. Hemost.* 42 (2016) 389–396. https://doi.org/10.1055/s-0036-1572327.

[81] Kakouros, N., J.J. Rade, A. Kourliouros, J.R. Resar, Platelet Function in Patients with Diabetes Mellitus: From a Theoretical to a Practical Perspective, *Int. J. Endocrinol.* 2011 (2011) 742719. https://doi.org/10.1155/2011/742719.

[82] Shacter, E., J.A. Williams, M. Lim, R.L. Levine, Differential susceptibility of plasma proteins to oxidative modification: Examination by western blot immunoassay, *Free Radic. Biol. Med.* 17 (1994) 429–437. https://doi.org/10.1016/0891-5849(94)90169-4.

[83] Maruhashi, JT., Soga, N. Fujimura, N. Idei, S. Mikami, Y. Iwamoto, M. Kajikawa, T. Matsumoto, Y. Kihara, K. Chayama, K. Noma, A. Nakashima, H. Tomiyama, B. Takase, A. Yamashina, Y. Higashi, Hyperbilirubinemia, Augmentation of

Endothelial Function, and Decrease in Oxidative Stress in Gilbert Syndrome, *Circulation.* 126 (2012) 598–603. https://doi.org/10.1161/CIRCULATIONAHA.112.105775.

[84] Cho, H.S., S.W. Lee, E.S. Kim, J. Shin, S.D. Moon, J.H. Han, B.Y. Cha, Serum bilirubin levels are inversely associated with PAI-1 and fibrinogen in Korean subjects, *Atherosclerosis.* 244 (2016) 204–210. https://doi.org/10.1016/j.atherosclerosis.2015.11.008.

[85] Ziberna, L., M. Martelanc, M. Franko, S. Passamonti, Bilirubin is an Endogenous Antioxidant in Human Vascular Endothelial Cells, *Sci. Rep.* 6 (2016) 29240. https://doi.org/10.1038/srep29240.

[86] Wufsus, A.R., N.E. Macera, K.B. Neeves, The Hydraulic Permeability of Blood Clots as a Function of Fibrin and Platelet Density, *Biophys. J.* 104 (2013) 1812–1823. https://doi.org/10.1016/j.bpj.2013.02.055.

[87] Hugenholtz, G.C.G., F. Macrae, J. Adelmeijer, S. Dulfer, R.J. Porte, T. Lisman, R.A.S. Ariëns, Procoagulant changes in fibrin clot structure in patients with cirrhosis are associated with oxidative modifications of fibrinogen, *J. Thromb.* Haemost. 14 (2016) 1054–1066. https://doi.org/10.1111/jth.13278.

[88] Tipping, E., B. Ketterer, L. Christodoulides, G. Enderby, Spectroscopic studies of the binding of bilirubin by ligandin and aminoazo-dye-binding protein A, *Biochem J* 157(1) (1976) 211-6.https://doi.org: 10.1042/bj1570211.

[89] Le Trong, R.E. Stenkamp, I., C. Ibarra, W.M. Atkins, E.T. Adman, 1.3-A resolution structure of human glutathione S-transferase with S-hexyl glutathione bound reveals possible extended ligandin binding site, *Proteins* 48(4) (2002) 618-27.https://doi.org: 10.1002/prot.10162.

[90] Sluis-Cremer, N., N.N. Naidoo, W.H. Kaplan, T.H. Manoharan, W.E. Fahl, H.W. Dirr, Determination of a binding site for a non-substrate ligand in mammalian cytosolic glutathione S-transferases by means of fluorescence-resonance energy transfer, *Eur J Biochem* 241(2) (1996) 484-8.https://doi.org: 10.1111/j.1432-1033.1996.00484.x.

[91] Bhargava, M.M., I. Listowsky, I.M. Arias, Ligandin. Bilirubin binding and glutathione-S-transferase activity are independent processes, *J Biol Chem* 253(12) (1978) 4112-5.https:// doi.org/10.1016/S0021-9258(17)34690-2

[92] Kamisaka, K., I. Listowsky, Z. Gatmaitan, I.M. Arias, Interactions of bilirubin and other ligands with ligandin, *Biochemistry* 14(10) (1975) 2175-80.https://doi.org: 10.1021/bi00681a021.

[93] Meuwissen, J.A., M. Zeegers, K.S. Srai, B. Ketterer, Effect of glutathione on the activity of bilirubin-binding proteins from rat liver cytosol [proceedings], *Biochem Soc Trans* 5(5) (1977) 1404-7.https://doi.org: 10.1042/bst0051404.

[94] Simons, P.C., D.L. Jagt, Bilirubin binding to human liver ligandin (glutathione S-transferase), *J Biol Chem* 255(10) (1980) 4740-4.https:// doi.org/10.1016/S0021-9258(19)85558-8

[95] van Raalte, D.H., M. Li, P.H. Pritchard, K.M. Wasan, Peroxisome proliferator-activated receptor (PPAR)-alpha: a pharmacological target with a promising future, *Pharm Res* 21(9) (2004) 1531-8.https://doi.org: 10.1023/b:pham.0000041444.06122.8d.

[96] Rakhshandehroo, M., B. Knoch, M. Muller, S. Kersten, Peroxisome proliferator-activated receptor alpha target genes, *PPAR Res* 2010 (2010).https://doi.org: 10.1155/2010/612089.

[97] Gordon, D.M., S.H. Hong, Z.A. Kipp, T.D. Hinds, Jr., Identification of Binding Regions of Bilirubin in the Ligand-Binding Pocket of the Peroxisome Proliferator-Activated Receptor-A (PPARalpha), *Molecules* 26(10) (2021).https://doi.org: 10.3390/molecules26102975.

[98] Stec, D.E., K. John, C.J. Trabbic, A. Luniwal, M.W. Hankins, J. Baum, T.D. Hinds, Jr., Bilirubin Binding to PPARalpha Inhibits Lipid Accumulation, *PLoS One* 11(4) (2016) e0153427.https://doi.org: 10.1371/journal.pone.0153427.

[99] Choi, S.H., K.E. Yun, H.J. Choi, Relationships between serum total bilirubin levels and metabolic syndrome in Korean adults, *Nutr Metab Cardiovasc Dis* 23(1) (2013) 31-7.https://doi.org: 10.1016/j.numecd.2011.03.001.

[100] Mustafa, M.F., B. Bano, Bilirubin binding with liver cystatin induced structural and functional changes, *J Fluoresc* 24(3) (2014) 967-74.https://doi.org: 10.1007/s10895-014-1381-4.

[101] Ochieng, J., G. Chaudhuri, Cystatin superfamily, *J Health Care Poor Underserved* 21(1 Suppl) (2010) 51-70.https://doi.org: 10.1353/hpu.0.0257.

[102] White, C.A., S. Ghazan-Shahi, M.A. Adams, beta-Trace protein: a marker of GFR and other biological pathways, *Am J Kidney Dis* 65(1) (2015) 131-46.https://doi.org: 10.1053/j.ajkd.2014.06.038.

[103] Kume, S., Y.H. Lee, Y. Miyamoto, H. Fukada, Y. Goto, T. Inui, Systematic interaction analysis of human lipocalin-type prostaglandin D synthase with small lipophilic ligands, *Biochem J* 446(2) (2012) 279-89.https://doi.org: 10.1042/BJ20120324.

[104] Beuckmann, C.T., M. Aoyagi, I. Okazaki, T. Hiroike, H. Toh, O. Hayaishi, Y. Urade, Binding of biliverdin, bilirubin, and thyroid hormones to lipocalin-type prostaglandin D synthase, *Biochemistry* 38(25) (1999) 8006-13.https://doi.org: 10.1021/bi990261p.

[105] Inoue, K., N. Yagi, Y. Urade, T. Inui, Compact packing of lipocalin-type prostaglandin D synthase induced by binding of lipophilic ligands, *J Biochem* 145(2) (2009) 169-75.https://doi.org: 10.1093/jb/mvn154.

[106] Inui, T., M. Mase, R. Shirota, M. Nagashima, T. Okada, Y. Urade, Lipocalin-type prostaglandin D synthase scavenges biliverdin in the cerebrospinal fluid of patients with aneurysmal subarachnoid hemorrhage, *J Cereb Blood Flow Metab* 34(9) (2014) 1558-67.https://doi.org: 10.1038/jcbfm.2014.127.

[107] Shoulders, M.D., R.T. Raines, Collagen structure and stability, *Annu Rev Biochem* 78 (2009) 929-58.https://doi.org: 10.1146/annurev.biochem.77.032207.120833.

[108] Bella, J. Collagen structure: new tricks from a very old dog, *Biochem J* 473(8) (2016) 1001-25.https://doi.org: 10.1042/BJ20151169.

[109] Kapoor, C.L. Interaction of bilirubin with reconstituted collagen fibrils, *Biochem J* 147(2) (1975) 199-203.https://doi.org: 10.1042/bj1470199.

[110] Usharani, N., G.C. Jayakumar, S.V. Kanth, J.R. Rao, B. Chandrasekaran, B.U. Nair, Identification of Bilirubin Binding Site in Type I Collagen, *International Journal*

of Peptide Research and Therapeutics 19(4) (2013) 357-364.https://doi.org: 10.1007/s10989-013-9359-7.

[111] Usharani, N., G.C. Jayakumar, J.R. Rao, B. Chandrasekaran, B.U. Nair, A microscopic evaluation of collagen-bilirubin interactions: in vitro surface phenomenon, *J Microsc* 253(2) (2014) 109-18.https://doi.org: 10.1111/jmi.12101.

[112] Nagarajan, U., J. Gladstone Christopher, B. Chandrasekaran, R.R. Jonnalagadda, U.N. Balachandran, K. Kohsaku, Studies on the molecular significance in the interaction of bilirubin with collagen, *Int J Biol Macromol* 61 (2013) 404-10.https://doi.org: 10.1016/j.ijbiomac.2013.08.001.

[113] Nagarajan, U., J.G. Christopher, R.R. Jonnalagadda, B. Chandrasekaran, U.N. Balachandran, Studies on the chemico-biological characteristics of bilirubin binding with collagen, *Mater Sci Eng C Mater Biol Appl* 33(8) (2013) 4965-71.https://doi.org: 10.1016/j.msec.2013.08.021.

[114] Blauer, G., G. Wagniere, Conformation of bilirubin and biliverdin in their complexes with serum albumin, *Journal of the American Chemical Society* 97(7) (1975) 1949-1954.https://doi: 10.1021/ja00840a057.

[115] Trull, F.R., O. Ibars, D.A. Lightner, Conformation inversion of bilirubin formed by reduction of the biliverdin-human serum albumin complex: evidence from circular dichroism, *Arch Biochem Biophys* 298(2) (1992) 710-4.https://doi.org: 10.1016/0003-9861(92)90470-h.

[116] Zsila, F., G. Mady, Biliverdin is the endogenous ligand of human serum alpha1-acid glycoprotein, *Biochem Biophys Res Commun* 372(3) (2008) 503-7.https://doi.org: 10.1016/j.bbrc.2008.05.090.

[117] Rosa, A., V.E. Pye, C. Graham, L. Muir, J. Seow, K.W. Ng, N.J. Cook, C. Rees-Spear, E. Parker, M.S. Dos Santos, C. Rosadas, A. Susana, H. Rhys, A. Nans, L. Masino, C. Roustan, E. Christodoulou, R. Ulferts, A.G. Wrobel, C.E. Short, M. Fertleman, R.W. Sanders, J. Heaney, M. Spyer, S. Kjaer, A. Riddell, M.H. Malim, R. Beale, J.I. MacRae, G.P. Taylor, E. Nastouli, M.J. van Gils, P.B. Rosenthal, M. Pizzato, M.O. McClure, R.S. Tedder, G. Kassiotis, L.E. McCoy, K.J. Doores, P. Cherepanov, SARS-CoV-2 can recruit a heme metabolite to evade antibody immunity, *Sci Adv* 7(22) (2021).https://doi.org: 10.1126/sciadv.abg7607.

[118] Minic, S.L., D. Stanic-Vucinic, J. Mihailovic, M. Krstic, M.R. Nikolic, T. Cirkovic Velickovic, Digestion by pepsin releases biologically active chromopeptides from C-phycocyanin, a blue-colored biliprotein of microalga Spirulina, *J Proteomics* 147 (2016) 132-139.https://doi.org: 10.1016/j.jprot.2016.03.043.

[119] Minic, S., D. Stanic-Vucinic, M. Radomirovic, M. Radibratovic, M. Milcic, M. Nikolic, T. Cirkovic Velickovic, Characterisation and effects of binding of food-derived bioactive phycocyanobilin to bovine serum albumin, *Food Chem* 239 (2018) 1090-1099.https://doi.org: 10.1016/j.foodchem.2017.07.066.

[120] Fernández-Rojas, B., J. Hernández-Juárez, J. Pedraza-Chaverri, Nutraceutical properties of phycocyanin, *Journal of Functional Foods* 11 (2014) 375-392.https://doi.org: 10.1016/j.jff.2014.10.011.

[121] Konickova, R., K. Vankova, J. Vanikova, K. Vanova, L. Muchova, I. Subhanova, M. Zadinova, J. Zelenka, A. Dvorak, M. Kolar, H. Strnad, S. Rimpelova, T. Ruml,

J.W. R, L. Vitek, Anti-cancer effects of blue-green alga Spirulina platensis, a natural source of bilirubin-like tetrapyrrolic compounds, *Ann Hepatol* 13(2) (2014) 273-83.https:// doi.org/10.1016/S1665-2681(19)30891-9

[122] Guo, W., M. Zeng, S. Zhu, S. Li, Y. Qian, H. Wu, Phycocyanin ameliorates mouse colitis via phycocyanobilin-dependent antioxidant and anti-inflammatory protection of the intestinal epithelial barrier, *Food Funct* (2022).https://doi.org: 10.1039/d1fo02970c.

[123] Strasky, Z., L. Zemankova, I. Nemeckova, J. Rathouska, R.J. Wong, L. Muchova, I. Subhanova, J. Vanikova, K. Vanova, L. Vitek, P. Nachtigal, Spirulina platensis and phycocyanobilin activate atheroprotective heme oxygenase-1: a possible implication for atherogenesis, *Food Funct* 4(11) (2013) 1586-94.https://doi.org: 10.1039/c3fo60230c.

[124] Pavon-Fuentes, N., J. Marin-Prida, A. Llopiz-Arzuaga, V. Falcon-Cama, R. Campos-Mojena, M. Cervantes-Llanos, B. Piniella-Matamoros, E. Penton-Arias, G. Penton-Rol, Phycocyanobilin reduces brain injury after endothelin-1- induced focal cerebral ischaemia, *Clin Exp Pharmacol Physiol* 47(3) (2020) 383-392.https://doi.org: 10.1111/1440-1681.13214.

[125] Zheng, J., T. Inoguchi, S. Sasaki, Y. Maeda, M.F. McCarty, M. Fujii, N. Ikeda, K. Kobayashi, N. Sonoda, R. Takayanagi, Phycocyanin and phycocyanobilin from Spirulina platensis protect against diabetic nephropathy by inhibiting oxidative stress, *Am J Physiol Regul Integr Comp Physiol* 304(2) (2013) R110-20.https://doi.org: 10.1152/ajpregu.00648.2011.

[126] Radibratovic, M., S. Minic, D. Stanic-Vucinic, M. Nikolic, M. Milcic, T. Cirkovic Velickovic, Stabilisation of Human Serum Albumin by the Binding of Phycocyanobilin, a Bioactive Chromophore of Blue-Green Alga Spirulina: Molecular Dynamics and Experimental Study, *PLoS One* 11(12) (2016) e0167973.https://doi.org: 10.1371/journal.pone.0167973.

[127] Minic, S., Bioactive properties of Spirulina-derived phycobiliproteins and phycobilins, *Biologia Serbica* 43(1) (2021) 32-42.https://dor.org: 10.5281/ zenodo.5512528.

[128] Gligorijevic, N., S. Minic, M. Radibratovic, V. Papadimitriou, O. Nedic, T.G. Sotiroudis, M.R. Nikolic, Nutraceutical phycocyanobilin binding to catalase protects the pigment from oxidation without affecting catalytic activity, *Spectrochim Acta A Mol Biomol Spectrosc* 251 (2021) 119483.https://doi.org: 10.1016/j.saa.2021.119483.

[129] Gligorijevic, N., M. Radomirovic, O. Nedic, M. Stojadinovic, U. Khulal, D. Stanic-Vucinic, T. Cirkovic Velickovic, Molecular Mechanisms of Possible Action of Phenolic Compounds in COVID-19 Protection and Prevention, *Int J Mol Sci* 22(22) (2021).https://doi.org: 10.3390/ijms222212385.

[130] Pendyala, B., A. Patras, In silico Screening of Food Bioactive Compounds to Predict Potential Inhibitors of COVID-19 Main protease (Mpro) and RNA-dependent RNA polymerase (RdRp), *ChemRxiv.* (2020). doi:10.26434/ chemrxiv.12051927.v2. This content is a preprint and has not been peer-reviewed. https://doi.org: 10.26434/chemrxiv.12051927.v2.

[131] Ghosh, S., S. Das, I. Ahmad, H. Patel, In silico validation of antiviral drugs obtained from marine sources as a potential target against SARS-CoV-2 Mpro, *Journal of the Indian Chemical Society* 98(12) (2021).https://doi.org: 10.1016/j.jics.2021.100272.

[132] Pendyala, B., A. Patras, C. Dash, Phycobilins as Potent Food Bioactive Broad-Spectrum Inhibitors Against Proteases of SARS-CoV-2 and Other Coronaviruses: A Preliminary Study, *Front Microbiol* 12 (2021) 645713.https://doi.org: 10.3389/fmicb.2021.645713.

[133] Petit, L., L. Vernes, J.P. Cadoret, Docking and in silico toxicity assessment of Arthrospira compounds as potential antiviral agents against SARS-CoV-2, *J Appl Phycol* 33(3) (2021) 1579-1602.https://doi.org: 10.1007/s10811-021-02372-9.

[134] Zhou, P., X.L. Yang, X.G. Wang, B. Hu, L. Zhang, W. Zhang, H.R. Si, Y. Zhu, B. Li, C.L. Huang, H.D. Chen, J. Chen, Y. Luo, H. Guo, R.D. Jiang, M.Q. Liu, Y. Chen, X.R. Shen, X. Wang, X.S. Zheng, K. Zhao, Q.J. Chen, F. Deng, L.L. Liu, B. Yan, F.X. Zhan, Y.Y. Wang, G.F. Xiao, Z.L. Shi, A pneumonia outbreak associated with a new coronavirus of probable bat origin, *Nature* 579(7798) (2020) 270-273.https://doi.org: 10.1038/s41586-020-2012-7.

Chapter 4

Overview of Panton-Valentine Leukocidin-Expressing *Staphylococcus Aureus* Pneumonia and Its Therapeutic Strategy

Jin-Wei Liu[1] and Wen-Liang Yu[2,3,*]

[1]Department of Medicine, Chi Mei Medical Center, Tainan, Taiwan
[2]Department of Intensive Care Medicine, Chi Mei Medical Center, Tainan, Taiwan
[3]Department of Medicine, School of Medicine, College of Medicine, Taipei Medical University, Taipei, Taiwan

Abstract

Panton Valentine leukocidin (PVL)-producing *Staphylococcus aureus* pneumonia is not a common but extremely life-threatening disease. Without immediate recognition and appropriate management, it can result in at least 75% mortality rate higher than PVL-negative strains. PVL is often found in community-associated methicillin-resistant *S. aureus* as well as community-associated methicillin-sensitive *S. aureus* isolates. The risk host factors of PVL-positive *S. aureus* pneumonia may include an original health individual, especially children and adolescents with a preceding respiratory virus infection, and the people close skin-to-skin contact with each other, like athletes, military trainees, jail inmates, etc. The pathogenesis of PVL may be attributed to the genes lukS-PV and lukF-PV, a pore-forming cytotoxin that can directly and indirectly cause polymorphonuclear leukocyte (PMN) destruction, resulting in severe leukopenia. Consistent with the PMN cytolytic activity of PVL, neutropenia is a frequent clinical finding associated with PVL-positive *S. aureus* necrotizing pneumonia. In addition, chest radiography may picture an initial mild pneumonia patch and can rapidly progress to

* Corresponding Author's Email: Yuleon_md@yahoo.com.tw.

In: Advances in Biology. Volume 1
Editor: Charles D. Grant
ISBN: 979-8-88697-201-6

diffuse multilobar consolidation with cavity formation around both lung fields. Hence, patients can present with acute respiratory distress syndrome and septic shock-like syndrome that require admission to an intensive care unit upon arrival of the emergent department or few days after admission. Currently, there is still no consensus for the management of this disease. Nevertheless, the detrimental effects of the infection are mainly caused by the production of PVL rather than the microorganism itself, therefore, an effective treatment should not only target at the eradication of *S. aureus* but additionally at eliminating the virulence of the toxin. We review the therapeutic strategies up to 2022 and recommend the following two methods. One is the combination of β-lactam antibiotics or glycopeptides with one kind of bacteriostatic antibiotics, acting as protein synthesis inhibitors, like clindamycin, linezolid, and tigecycline. The other is the addition of a combination of intravenous linezolid with intravenous clindamycin and rifampicin to the initial empirical antibiotic regimens if strongly suspected PVL-positive *S. aureus* pneumonia. Meanwhile, adjunctive therapy like intravenous immunoglobulin (IVIG) is considered to inactivate the toxin as soon as possible. IVIG is believed to neutralize PVL pore formation and the cytopathic effects of PVL, bringing beneficial benefits to the prognosis. For those critical patients with unstable vital signs, the invasive life-supportive methods like mechanical ventilation or even the use of extracorporeal membrane oxygenation should be initiated without delay. Moreover, the cytokine adsorption therapy has emerged as a valuable adjuvant therapy and it can rapidly reduce the inflammatory status, thus lowering the need or dosage of vasopressor agents. In conclusion, due to the considerable lethality of PVL-expression *S. aureus* pneumonia, early identification of the PVL gene is so crucial that physicians can take the direct therapeutic strategies to enhance the success of saving patient's life and perhaps could avoid further invasive life-sustained treatment.

Keywords: clindamycin, linezolid, leukocidin, pneumonia, PVL, *staphylococcus aureus*, tigecycline

1. Introduction

Staphylococcus aureus is a major cause of infectious disease. However, it is thought to be the colonies in the nasal mucosa and might not lead to illness [1]. On one hand, following its colonization, subsequent infection may occur if they undergo nasal invasive procedure or there is impaired mucosal barrier [2]. On the other hand, it can be taken for the normal flora around our skin but can also be regarded as the pathogen when there are skin and soft tissue infections (SSTIs), ranging from minor and self-limited skin infections to invasive and life-threatening diseases [3].

Unfortunately, *S. aureus* can spread out to anywhere in our human body causing bacteremia if the focal infections were untreated properly [4]. Moreover, due to its strong ability of adhesion and internalization to host cells, it can result in many intractable diseases, like sepsis, infective endocarditis, septic meningoencephalitis, deep-seated muscular abscess, septic arthritis, spondylodiscitis, and osteomyelitis [5]. Herein, we focus on the issue of pneumonia caused by *S. aureus*.

Panton–Valentine leukocidin (PVL) is an exotoxin secreted by *S. aureus* and could also increase the virulence of this bacteria. It was thought to be the culprit of severe infections caused by *S. aureus*, such as necrotizing pneumonia and necrotizing fasciitis [6]. In a study comparing eight retrospective and eight prospective cases of PVL-positive *S. aureus* pneumonia with 36 cases of PVL-negative *S. aureus* pneumonia, the 48-hour survival rate after admission was 63% for the PVL-positive patients and 94% for PVL-negative individuals ($p = 0.007$), suggesting the close link between the existence of PVL expression and the high mortality rate [7].

The rapid emergence of methicillin-resistant *S. aureus* (MRSA) is a considerable global health issue in these years. It was widely recognized that the widespread use of antibiotics has generated selective pressures that have driven the appearance of resistant strains [8]. Therefore, MRSA infections have traditionally been hospital-associated (HA-MRSA). However, in recent years, a steadily increasing number of MRSA infections have been due to the community-associated MRSA (CA-MRSA) strains [9]. It cannot be explained by the above reason completely due to the low-level use of antimicrobial agents.

Transduction of staphylococcal cassette chromosome *mec* (SCCmec) genetic elements, responsible for methicillin resistance, mediates a novel penicillin binding protein designated PBP2a, possessing lower affinity for the β-lactam antibiotics than the native PBP2 enzyme, thus becoming the MRSA strains of public health concern [10]. Bacteriophage can transfer SCCmec gene to another strain of *S. aureus* and it can also integrate the gene in charge of the production of PVL by *S. aureus* [11]. Hence, not only can be found in the community-associated methicillin-sensitive *S. aureus* (CA-MSSA) isolates, PVL can often be found in majority of CA-MRSA rather than in the HA-MRSA isolates [12]. Although the extreme virulence and high mortality of PVL-positive *S. aureus*, PVL is less often present in HA-MRSA isolates, which is usually associated with asymptomatic nasal colonization [13].

2. Epidemiology

In the early 1960s, the adaptive power of *S. aureus* to antibiotics led to the emergence of MRSA [14]. However, from the late 1990s, CA-MRSA clones emerged worldwide and are often associated with the presence of the toxin PVL [15]. A study in Western Nepal, PVL was highly prevalent among CA-MRSA (90.4%), while PVL was detected only in 7.1% HA-MRSA strains [13].

Infections caused by MRSA can be endemic or epidemic based on its clonal structure and sequence typing around the world. Within most hospitals in Asia, an estimated proportion of HA-MRSA ranged from 28% (in Hong Kong and Indonesia) to >70% (in Korea) among all clinical *S. aureus* isolates in the early 2010s [16]. While in Western countries, MRSA incidence among *S. aureus* also varies country by country with a low frequency (around 1%) in some countries in Europe (e.g., the Netherlands) and a high frequency (> 60%) in the United States [17, 18].

On the other hand, as to CA-MRSA infection, the incidence of CA-MRSA in Asia varies markedly among countries, too. An ANSORP study demonstrated the proportion of CA-MRSA in *S. aureus* infections varied by country, such as Sri Lanka (38.8%), Taiwan (34.8%), the Philippines (30.1%), Vietnam (30.1%), Korea (15.6%), Hong Kong (8.5%), India (4.3%), and Thailand (2.5%) [19]. In recent years, the number of adult patients with CA-MRSA infections increased over time, and the disease was associated with more cases of necrotizing pneumonia and cutaneous abscess [20].

We have already known the prevalence of PVL predominantly exists among CA-MRSA rather than HA-MRSA. However, not only can PVL be carried by CA-MRSA but it can also be presented around CA-MSSA [21]. For example, no difference was found between the prevalence rate of PVL in MRSA (57.1%) and in MSSA (61.5%) among patients seen at an urban Gambian hospital. However, the phenomenon may be attributed to the low prevalence of MRSA rates within the community of Gambia [22]. If we look inside the area with high prevalence of MRSA, like in the United States, there is a prominently higher incidence of PVL in MRSA than MSSA. (48.1% versus 11.5% respectively) [23].

Even if we had thought an association between PVL and CA-MRSA infection deeply rooted over the past decades, however, MRSA clones have spread between the community and hospitals during recent years, leading to the obscure and challenging distinction between HA-MRSA and CA-MRSA [19]. There is more and more evidence suggesting the successful epidemic of

PVL-positive MRSA clones entering the hospital and causing nosocomial infections [24]. For example, PVL-positive MRSA isolates were more common in inpatients (69.2%), especially isolates from deep tracheal aspirate (30.8%) and abscess/wound (23.1%) in a university hospital in Northern Cyprus [24].

Despite the complicated relationship of PVL between CA-MRSA and HA-MRSA, the risk factors of these two host groups were exactly divergent. The risk hosts of HA-MRSA included hemodialysis, surgery, residence in a long-term-care facility or hospitalization during the previous year, the presence of an indwelling catheter or percutaneous device at the time of culture, or previous isolation of MRSA [25]. However, on the contrary, CA-MRSA infections mainly occur in healthy individuals, especially children and adolescents, through skin-to-skin contact [26]. Reflecting this, risk host factors for CA-MRSA infection includes university and school students, adolescent athletes, military trainees, jail inmates, men who have sex with men, members of an infected family, hurricane evacuees, and people with tattoos [27, 28].

3. Clinical Manifestations

Many coagulase-positive and coagulase-negative *Staphylococci* are not causing illness as they are normal microbiota of the nasal cavity [29]. For example, *S. aureus* harmlessly colonizes the skin and mucosa of about 20% to 30% averagely in healthy adults and children [30, 31]. Nonetheless, they may invade our body and result in various diseases. Additionally, PVL-positive *S. aureus* is often considered a true pathogen [32], which has particularly strong association with more severe clinical scenarios [33], such as severe necrotizing pneumonia in MRSA or MSSA, especially in community [34, 35].

As to primary community-acquired pneumonia (CAP) caused by PVL-positive *S. aureus*, severe bilateral pneumonia could occur in an originally healthy adult patient who presents with acute respiratory distress syndrome and septic shock-like syndrome that require admission to an intensive care unit [36]. Concurrent or recent acute respiratory virus infections, mostly influenza and parainfluenza, are common but not always present [37]. Patients who developed severe necrotizing pneumonia caused by PVL-positive *S. aureus* commonly have a preceding "flu-like" illness, featuring like fever, headache, arthralgia, myalgia, chest pain, tachycardia, tachypnea, shortness of breath, nausea, vomiting, diarrhea, sore throat, rhinorrhea, nasal congestion, productive or dry cough, and even hemoptysis [38, 39, 40, 41]. Unstable vital signs, including hypotension and respiratory distress may not appear upon arrival of the emergent department but it can be fulminantly progressive and even deteriorate to need intubation with mechanical ventilation support just within short days of the clinical course [42, 43].

In the 21st century, a global outbreak of SARS-CoV-2 in 2020 makes every person worldwide panicky and anxious. It absolutely alters our life and even limits our daily activity. Likewise, epidemiology and clinical characteristics of relative diseases were also influenced by this new Coronavirus. For example, as for PVL-expression necrotizing pneumonia, there is a growing number indicating the co-infection with COVID-19, similarly to the influenza virus, damaging the bronchial epithelium and facilitating invasive infection of virulent *S. aureus*, which may not only colonize the respiratory tract [44, 45]. The association between influenza and severe PVL-expression *S. aureus* infection is well described, however, it needs more studies for establishing the relationship between SARS-CoV-2 infection and subsequent fatal PVL-expression *S. aureus* necrotizing pneumonia.

Severe necrotizing pneumonia caused by PVL-expression *S. aureus* can not only originate from the pulmonary site secondary to antecedent respiratory virus infection, but it can also come from other body site of *S. aureus* infections, for example, secondary to SSTIs [46, 47]. Even if PVL-positive MRSA strains were mainly found in SSTIs [48], it can also contribute to kinds of diseases, such as bacteremia, infective endocarditis, septic arthritis, osteomyelitis, etc. [5]. All the above diseases can be the beginning clinical manifestation of PVL-expression *S. aureus* necrotizing pneumonia.

4. Pathogenesis

With a 2.8Mb genome composed of 75% core chromosome and 25% accessory component, *S. aureus* is a formidable bacterial pathogen of human infections. The core is highly conserved among strains while its accessory component is considerably variable [49]. Nevertheless, much of the genetic diversity of MRSA and other pathogens occur within the accessory genome, consisting of mobile genetic elements (MGEs), such as pathogenicity islands, bacteriophages, chromosomal cassettes, transposons, and plasmids, acquired by horizontal transfer between strains, responsible for mediating virulence, immune evasion, and antibiotic resistance [50, 51].

The pathogenesis of PVL may be attributed to the genes lukS-PV and lukF-PV located in the MGEs, encoded in prophage Sa2int. PVL is a pore-forming cytotoxin that can directly cause leukocyte destruction, resulting in severe leukopenia. Indirectly, it can initiate the apoptosis pathway via destructing the mitochondria of leukocyte with the release of reactive oxygen species (ROS) and inflammatory mediator, like caspase-3, caspase-9, cytochrome c and second mitochondria-derived activator of caspase/direct inhibitor of apoptosis-binding protein with low pI (Smac/DIABLO) resulting in leukopenia and soft tissue necrosis, even necrotizing pneumonia [52, 53, 54, 55].

Whereas the cytotoxic action of PVL on human neutrophils is already well established, the toxicity of PVL can also influence other sensitive cell types, such as monocytes and macrophages but not to lymphocytes. Specifically, LukS-PV, the critical genetic factor for PVL-induced cytotoxicity, was identified as the most important part rather than lukF-PV [56].

PVL consists of PVL-S (284 amino acids, 32 kDa) and PVL-F (301 amino acids, 34 kDa), working together to destroy the leukocyte in our human body [57, 58]. Generally, the S-component takes charge of binding to the polymorphonuclear leukocytes (PMNs) membrane with high affinity while the F-component plays a role as a toxic component [59]. They can assemble into the channel complex on cell membranes and induce necrosis of PMN and monocytes within 30 minutes, and induce apoptosis of PMNs within 4 hours. Essentially, they need each other to obtain the effect of cytotoxicity, hence, neither PVL-S nor PVL-F alone induces necrosis even at more than ten times higher concentrations [60].

Not only PVL, there is a wide range of virulence factors being secreted by *S. aureus*, including toxins (hemolysins, toxic shock syndrome toxin, enterotoxin, and exfoliative toxin), immune-evasive surface factors (for example, capsule and protein A), and enzymes that promote tissue invasion (for example, hyaluronidase) [61].

Moreover, a synergistic effect of PVL and other virulence factors had been demonstrated. For instance, *S. aureus* can produce staphylococcal protein A (SpA), a surface protein, found in the cell wall of the bacteria or freely secreted into the extracellular environment. It has the capability to bind immunoglobulins, most notably immunoglobulin G (IgG). Through binding the heavy chain within the Fc region of most IgG and the Fab region of the B-cell receptor, it can impair the immune response of the human body and potentially promote immune evasion, including the inhibition of phagocytosis and opsonization. In addition, SpA is not only able to suppress the immunity of host but can facilitate the invasiveness of *S. aureus* and the severity of staphylococcal pneumonia [62, 63]. Furthermore, PVL can increase the expression of SpA, leading to a more toxic and more lethal clinical situation [64].

Alpha-hemolysin (α-toxin), just like PVL, is also a pore-forming exotoxin produced by *S. aureus*, causing the destruction of erythrocytes and several other types of cells. Also, it can wreck the PMNs through the stimulation of arachidonic acid pathway inside and followed pulmonary thromboxane generation, that can contribute to the pathogenesis of pulmonary hypertension and acute respiratory failure. Besides, α-toxin triggers the oversecretion of IL-1α and IL-6, participating in the mechanism of inflammation and known to induce organ injury. More terribly, the synergy effect between α-toxin and PVL together can potentially result in necrotising vasculitis with massive lung areas of infarction and hemorrhage [65, 66, 67, 68].

Necrotizing pneumonia caused by PVL-positive *S. aureus* can not only owe to the violent cytotoxicity of this virulence factor. Prior respiratory virus infection is also the major reason to explain the mechanism. Influenza virus is known to reduce secretion clearance and facilitate bacteria adhesion by hindering phagocytic killing and damaging the bronchial epithelium. It also precipitates the migration of immune cells to lung tissues, and thus the rapid killing of these cells by PVL brings the release of inflammatory mediators, which may promote the disease development of necrotizing pneumonia and even the phenomenon of cytokine storm [69].

5. Diagnosis

When there is clinical suspicion of the presence of PVL-expression *S. aureus* pneumonia, immediate laboratory diagnosis and image study are necessary. However, before carrying out the tools for diagnosis, we can initially make an assessment for whether there is the presence of 'typical' clinical scenarios or based on the epidemiology and clinical manifestations of PVL–positive *S. aureus* pneumonia [70]. Nonetheless, we still need further examination and blood tests to help us establish a correct diagnosis.

5.1. Molecular Diagnosis

Firstly, we can easily refer the sample to Staphylococcus Reference Laboratory for toxin gene profiling, including PVL testing. This is a polymerase chain reaction (PCR)-based assay that can be completed within a working day [70]. Moreover, we can get further laboratory investigation if utilizing other techniques used to analyze the genome of *S. aureus*, like multilocus sequence typing (MLST), pulsed-field gel electrophoresis, Spa typing, and other special forms of PCR [51, 71].

Especially MLST, a DNA sequence-based typing method with a public database accessible online for interlaboratory data comparisons, has highly discriminatory with the pandemic strains clearly represented by numerous sequence types (STs) and clonal complex. Additionally, it is well applied for epidemiologic investigations and evolutionary relatedness [72, 73]. For example, according to the literature in Taiwan, sequence type 188 was predominant in MSSA infections, while ST59 was the common type in MSSA and MRSA, and PVL-positive ST59 *S. aureus* was a successful clonal lineage in both MRSA and MSSA [74].

5.2. Laboratory Testing

Consistent with the PMN cytolytic activity of PVL, neutropenia is a frequent clinical finding associated with PVL-positive *S. aureus* necrotizing pneumonia. In contrast to the classic finding of leukopenia, the white blood cell count (WBC) in some PVL-positive *S. aureus* infections may slightly increase with or without bandemia or even presents as a leukemoid reaction.

Positive *S. aureus* blood cultures may appear, which are comparatively unusual in PVL-associated pneumonia. Gram staining of the sputum may reveal sheets of staphylococci, and possibly with a paucity of neutrophils with some red cells [75].

Toxic shock syndrome and myositis should be suspected if the creatine kinase is significantly raised. C-reactive protein (CRP) has a high probability to increase up to an extreme level, reflecting the gross tissue devastation [76, 77, 78].

Inflammatory cytokines, such as Interleukin-2 (IL-2), Interleukin-6 (IL-6) and Interleukin-8 (IL-8), may be elevated to an extraordinary level amazingly as cytokine-release syndrome [79, 80]. If disease still progresses, disseminated intravascular coagulation can occur and then increase the mortality [42]. Other than the appearance of cytokine storm, due to the adverse effect of SpA, measured serum IgG may be below the normal range of 540~1822 mg/dl.

According to the epidemiologic tendency of preceding respiratory virus infection around PVL-positive *S. aureus* pneumonia, the Filmarray respiratory panel (FARP) can reliably and rapidly identify the existence of common respiratory virus infections if they still coexisted [81].

5.3. Image Study

The most convenient and accessible image study for the evaluation of PVL-positive *S. aureus* pneumonia is absolutely chest radiography. The commonest radiological presentation of staphylococcal pneumonia is the appearance of several small rounded areas of consolidation that usually cavitate within 96 hours on serial radiographs and coalescence of small cavities resulting in pneumatocele, fistulae, empyema, or necrosis as the trouble complications [82].

Therefore, physicians should be careful when single or multiple opacities <3 cm diameter, especially with cavitation is found in chest radiography, suggestive of staphylococcal infection. The incidence of complicated pneumonia is far higher than non PVL-producing staphylococcal pneumonia [75].

Different from non PVL-producing staphylococcal pneumonia, PVL-positive *S. aureus* pneumonia could be completely normal or with minimal findings within chest radiography in the first 24–48 h. However, due to its unique clinical manifestation, the development of radiological change is thereafter very rapid. It can quickly progress from one lobar pneumonia to one lung white-out lesion and even to multilobular alveolar infiltration within one day. The worst, diffuse multilobar opacities followed by cavity formation around both lung fields can be like the pattern of acute respiratory distress syndrome (ARDS) [75, 83, 84, 85].

PVL toxin-related *S. aureus* necrotizing pneumonia, if untreated properly, can bother clinicians with the complications including pneumatoceles, pleural effusion, empyema, lung abscess and diaphragmatic paralysis, building an obstacle difficult to overcome [39, 85, 86]. Due to the catastrophic destruction of necrotizing pneumonia and necrotizing tracheobronchitis caused by PVL-positive *S. aureus* pneumonia, it is possible to see the emergence of tension pneumothorax or hemothorax, complicating the clinical course [39, 87, 88].

PVL-positive *S. aureus* pneumonia can be a primary lung infection or be secondary to hematogenous spreading by *S. aureus* bacteremia. For persistent *S. aureus* bacteremia, echocardiography is usually the routine examination to exclude the possibility of infective endocarditis. Meanwhile, for severe infections caused by PVL-positive *S. aureus* pneumonia, there is high probability to see cardiac global hypokinesis as septic cardiomyopathy. Additionally, we can also arrange computed tomography or magnetic resonance imaging to evaluate the definite severity of necrotic and liquefied lungs, particularly with cystic changes or the possible coincidence of pulmonary and even systemic septic embolism events, like cerebral emboli, lower extremities emboli, etc. [47, 89].

Bronchoscopy, an invasive procedure that lets physicians look at the lung and airway lesions, is usually performed in certain situations. For the suspicion of PVL-positive *S. aureus* pneumonia, obtaining samples of mucus or tissue via bronchoalveolar lavage can be helpful to quickly identify the pathogen and to analyze the form of pneumonia [39, 40].

6. Treatment

6.1. Antimicrobial Therapy

There are more and more cases of PVL-positive *S. aureus* pneumonia being reported. However, currently, there is no guideline for the management of this disease. Take antibiotic therapy for example, there are various opinions from global experts. All in all, the consensus of combining at least two antibiotics with different mechanisms was the mainstream. The use of antibiotics with β-lactam alone for treating PVL-positive *S. aureus* may increase the secretion of ribosomal exoprotein toxins resulting in more inflammatory response due to the enhancement of PVL and hemolysin release, especially oxacillin with sub-inhibitory concentrations inside the tissue [46, 47, 75, 90, 91]. Hence, the combination with one of the antibiotics acting on protein synthesis is very important to avoid the increased inflammatory response causing more violent injury of lung tissue. Traditionally, the most common choice is clindamycin, a bacteriostatic agent possessing the strong inhibitory ability for reducing the production of PVL and the inflammation feature. Also, the dose of clindamycin can be titrated up to 900 milligram every eight hours if poor response [42, 75, 92, 93].

However, β-lactams can be used as the bactericidal agents for treating MSSA infections while it is useless when confronting MRSA infections. This time, the role of glycopeptides, like vancomycin and teicoplanin are essential for wiping out the bacteria. Fortunately, glycopeptides have no effect on exotoxin formation not leading to the oversecretion of PVL. Nonetheless, due to its inadequate capability of lung penetration, its therapeutic efficacy remains controversial [43, 87]. On the other hand, other bactericidal agents for treating MRSA infections, such as fosfomycin, had been reported to successfully treat PVL-positive *S. aureus* pneumonia combined with one of the bacteriostatic agents [38].

As to the use of fluoroquinolone, there is no study to compare the efficacy with other antibiotics. However, its susceptibility to MSSA or MRSA is equivocal and there are also many cases demonstrating the failure to use it as the empiric antibiotics. Thus, it is not recommended to treat PVL-positive *S. aureus* pneumonia until the result of ultimate antimicrobial susceptibility test and there is no other choice.

In general, clindamycin is sensitive to CA-MRSA while it is almost resistant to HA-MRSA. However, an ambiguous distinction between CA-MRSA and HA-MRSA has emerged in these years due to the intercommunication between community and hospital [19]. Therefore, it is not recommended when there is a risk factor of hospital-acquired infection or in the community with high resistance rate to clindamycin.

Instead, clindamycin can be replaced with linezolid, tigecycline, or fusidic acid as protein synthesis inhibitors if worried about the resistance of antibiotics and the results are good. [75, 90, 94, 95, 96]. In addition, there is one case report revealing good response of tigecycline used alone first, achieving the success of decreased CRP level, indicating the well control of inflammatory response [38]. Rarely, there are few reports showing the successful application of aminoglycoside, macrolide, sulfamethoxazole/ trimethoprim, or tetracycline for treating PVL-positive *S. aureus* pneumonia.

Daptomycin, an extremely rapidly bactericidal antibiotic, unfortunately, is not suitable for treating PVL-positive *S. aureus* pneumonia because of the effect of surfactant inactivation on the drug in the lung tissue.

Since the detrimental effects of the infection are mainly caused by the production of PVL rather than the microorganism itself, an effective treatment should not only target at the eradication of *S. aureus* but additionally at eliminating the virulence of the toxins [43]. Some studies even emphasized the dominant position of reducing toxin over killing bacteria itself. Therefore, the collaboration of two kinds of protein synthesis inhibitors as dual therapy is also recommended in some guidelines. The most evident and necessary active regimens for disrupting toxin production is the combination of linezolid plus clindamycin [39, 40, 43]. Apart from the statements above, British

Thoracic Society guidelines published in 2015 had mentioned antibiotic therapy should include the addition of a combination of intravenous linezolid with intravenous clindamycin and rifampicin to the initial empirical antibiotic regimens if strongly suspected PVL-positive *S. aureus* necrotizing pneumonia until it is either confirmed or excluded. Taking linezolid, clindamycin, and rifampicin together as triple therapy to treat PVL-positive *S. aureus* pneumonia can be more beneficial than dual therapy. [46, 80, 97, 98]. However, this combination regimen increases the loading of liver metabolism and potential toxicity.

Specially, there is a study showing good results and efficacy of high dose tigecycline (that is, 100 mg every 12 hours after a 200 mg loading dose) improving the clinical outcome of critically ill patients with multidrug-resistant pathogens infections [99]. Prospectively, maybe high dose tigecycline is a good choice for treating PVL-positive *S. aureus* necrotizing pneumonia and further investigation is warranted in the future.

With the passage of time, carbapenems, newly developed cephalosporins and glycopeptides, still need more studies and case reports to clarify the issue of whether these antibiotics can be effective or not. In addition, up to now, there is still a lack of trials to prove which antibiotic is superior to another [97, 100].

6.2. Adjunctive Therapy

Early diagnosis of PVL-positive *S. aureus* pneumonia is essential due to the useless effect of typical empiric antimicrobial therapy for CAP. Also, we emphasized that antibiotic regimens should cover MRSA as a significant proportion of PVL-positive strains being CA-MRSA [23]. As the PVL toxin will continue to be active and destructive even if the bacteria have already been successfully killed. Accordingly, adjunctive therapy should be initiated to inactivate the toxin itself as soon as possible [98].

Intravenous immunoglobulin (IVIG) is believed to neutralize PVL pore formation and the cytopathic effects of PVL, and thus could be an adjunctive agent helpful to the patients. When treating fatal PVL-positive *S. aureus* pneumonia, not only selecting appropriate antibiotics but also combining them with IVIG without delay may be beneficial to the prognosis. The appropriate dose of IVIG is still not determined by convincible research. However, it is recommended with the dose of 2 g/kg and can be repeated after 48 hours if there is still evidence of sepsis [43, 87, 101].

It is not uncommon to see unstable vital signs among patients suffering from PVL-positive *S. aureus* pneumonia. This time, hydrocortisone administration can be added to the therapeutic regimen because of septic shock. For patients with leukopenia found in the blood tests, an adverse effect of PVL toxin, granulocyte colony-stimulating factor can also be given to the patients. By the way, due to the close link with preceding influenza virus infection, drugs such as oseltamivir used to treat influenza A and influenza B can be simultaneously prescribed to the patients if confirmed influenza diagnosis [39, 43].

Activated protein C (APC) has been characterized with the direct cytoprotective effects, including anti-inflammatory, anti-apoptotic activities, and endothelial barrier stabilization. Anecdotal reports indicate a possible role that recombinant APC therapy can be used to treat severe sepsis, such as PVL-positive *S. aureus* pneumonia. However, it conveys an increased risk of serious bleeding complications due to APC anticoagulant activity. Unfortunately, it is relatively easy to see the adverse outcome of pulmonary hemorrhage in PVL-positive *S. aureus* pneumonia due to its violently destructive power. Hence, it is still not recommended by most experts for the awareness of the risk of precipitating further hemorrhage [36, 75, 102, 103]. Whereas, there is a novel agent, recombinant APC variants, providing relatively enhanced levels of either cytoprotective or anticoagulant activities but with reducing bleeding risk, is a promising product for treating severe sepsis. Even so, there is still no available data to support its application [104, 105].

PVL-positive *S. aureus* pneumonia is a potentially life-threatening disease leading to ARDS-associated respiratory failure needing ventilatory support. However, it is possible to meet the clinical scenario that refractory hypoxemia with persistent carbon dioxide (CO2) retention may occur despite the setting of the ventilator. This time, only the initiation of extracorporeal membrane oxygenation (ECMO) can save the patient's life. Although the use of ECMO for PVL-positive *S. aureus* pneumonia had largely been unsuccessful, there is more and more case reports showing the encouraging experience, suggesting that ECMO may be a useful therapeutic tool for rescuing critical patients suffering from this disease [88, 103, 106].

The choice of venoarterial ECMO (VA-ECMO) or venovenous ECMO (VV-ECMO) depends on the stability of hemodynamics and the presence of septic cardiomyopathy. For example, VA-ECMO is proper when there is left ventricular dysfunction associated with right ventricular dilatation (Figure 1). However, if available, VV-ECMO is better than VA-ECMO due to the excellent oxygenation and the normal physiology. Therefore, VA-ECMO should be shifted to VV-ECMO as quickly as improved clinical conditions [107].

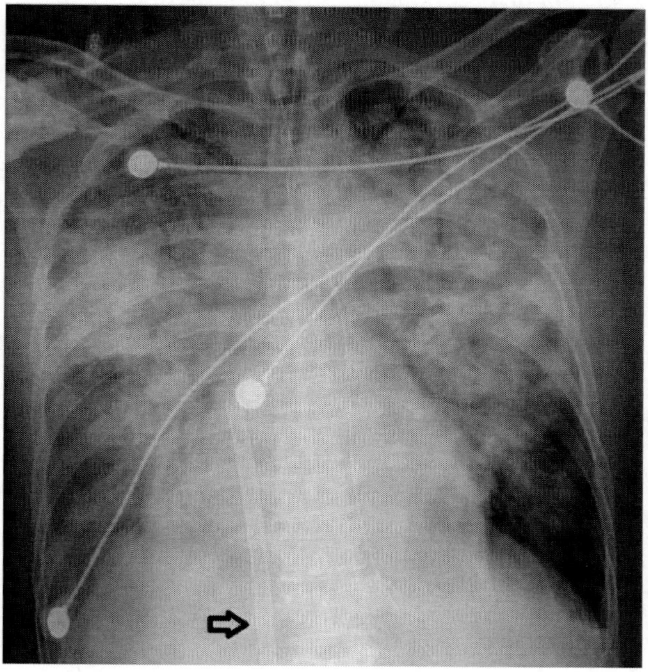

Figure 1. A case of PVL-positive CA-MRSA pneumonia with consolidation in bilateral lung fields, requiring VA-ECMO placement (arrow) to maintain adequate oxygenation during unstable hemodynamic status.

ECMO can be recognized as an effective technique to treat acute respiratory failure, especially for patients with severe pulmonary conditions thought to be pathophysiologically reversible. The advantage of ECMO is the permission for the lung-protective ventilation, avoiding barotrauma or ventilation-induced lung injury and the allowance for lessening the burden of heart. We can also acquire adequate time to endeavor for the elimination of PVL toxin and eradication of *S. aureus* [106, 108].

Although the pros and cons, the benefit, and the balance of economic cost of the use of ECMO are still questioned, we still recommend the initiation of ECMO without hesitancy if clinical conditions progressed beyond our expectancy. Septic shock was traditionally seen as a contraindication for ECMO; however, for those suffering from rapidly progressive cardiovascular collapse, it can be recognized as a hope to reverse the illness and prevent further multiple organ failure [109].

Strikingly, we may have an encounter with persistent hypotension despite the use of extracorporeal life support. This time, severe vasoplegia caused by the cytokine storm should be suspected. Due to extreme distributive shock status, strong inotropic agents may be added, however, it just procrastinates the death of the patient. PVL-positive *S. aureus* pneumonia can cause clinical conditions involving various cytokine cascades; hence, cytokine adsorption therapy has emerged as a valuable adjuvant therapy. It can reduce the inflammatory status rapidly from the blood by removing inflammatory mediators. Afterwards, we can gradually decrease the need of vasopressor dose [110, 111, 112, 113, 114, 115, 116]. There is a randomized, controlled pilot study demonstrating the success of early application of extracorporeal cytokine adsorption therapy in septic shock to be safe with significant effects [117]. However, it remains to be explored whether cytokine adsorption therapy can directly get rid of the PVL toxin.

While admission, it is liable to see pulmonary complications due to fragile lungs destroyed by PVL-positive *S. aureus.* Immediate invasive procedures like thoracic drainage system or chest surgical intervention must be considered once the presence of pneumothorax, empyema, or lung abscess, etc. [85]. Less common complications such as systemic emboli, ischemia limb or ischemial bowel and infective endocarditis deserve physicians to pay more attention and corresponding management should be implemented promptly.

Except for the therapeutic strategies mentioned above, there is an innovative way for dealing with PVL-positive *S. aureus* pneumonia. Targeting NLRP3 inflammasome and C5aR1, the receptor of PVL toxin in human-induced pluripotent stem cell-derived macrophages (hiPSC-dMs) can protect hiPSC-dMs away from PVL cytotoxicity [118]. However, it is still under laboratory hypothesis, thus, we remain neutral due to the low evidence level now.

The patient's ultimate survival and recovery depend on the extraordinary interplay and the coordinated approach of numerous multidisciplinary specialists. Early recognition of the PVL-positive *S. aureus* infections is so imperative that adequate antibiotic regimens with combining evident adjuvants can prevent the disease from progressing to mechanical ventilation or even the use of ECMO. However, with aggressive and judicious management, it is not an incurable deadly disease.

7. Clinical Outcome

PVL-expression *S. aureus* infection usually results in poorer prognosis and higher mortality than PVL-negative strains with risk of death 1.56-fold (95% confidence interval) and increased risk of sepsis by 5.51 times [48, 119, 120]. Interestingly, there are some studies challenging this traditional view. For instance, a systematic review and meta-analysis published in 2013 did not identify the evidence to support a link between PVL genes and poor clinical outcome [32]. However, this result might be indicative of underlying population differences in terms of age and comorbidities, because elderly people are more likely to be infected with PVL-negative MRSA than are younger adults. In addition, the role of PVL is controversial while the presence of PVL was not the primary determinant of clinical outcome in patients with complicated SSTIs [121].

Despite the debate of the correlation between PVL-positive *S. aureus* infection and its poor prognosis, it is unanimous that PVL-positive *S. aureus* inducing necrotizing pneumonia can contribute to at least 75% mortality rate higher than PVL-negative strains with a significant disparity [102, 122, 123]. Patients may pass away finally even under ventilator support and extracorporeal circulation.

Nonetheless the presence of methicillin resistance in *S. aureus* is not a predictor of severity in community-acquired *S. aureus* necrotizing pneumonia and there is no difference in the mortality rates of PVL-positive CAP between MRSA and MSSA [124, 125].

If treated properly and immediately, patients may survive from PVL-expression *S. aureus* pneumonia; however, they may suffer from a period needing home oxygen therapy or continuous positive airway pressure due to severely reduced lung capacity with a forced vital capacity and destructed pulmonary constitution with deformed airway and trachea structure [42].

8. Conclusion

Necrotizing pneumonia caused by PVL–secreting *S. aureus* is not a common but extremely life-threatening disease. Without instant identification and adequate management, its mortality can reach up to 75%. PVL no doubt plays an indispensable role in the pathogenesis of this disease. A high degree of awareness is necessary for the early detection of PVL-positive *S. aureus* pneumonia and physicians should rapidly decide the most appropriate strategies to contain the disease.

Currently, there is no single best way to treat PVL-positive *S. aureus* pneumonia; however, a multidisciplinary approach appears to offer the best prognosis. As the identification of the toxin is not always feasible or quickly performed, empirical choice of antimicrobials is of paramount importance. Early application of a combination of intravenous linezolid with intravenous clindamycin and rifampicin to the initial empirical antibiotic regimens used to treat severe CAP caused by CA-MRSA can probably enhance the success rate to save patient's life and perhaps could avoid further invasive life-sustained treatment.

Additionally, adjuvants therapy such as IVIG should be given to the patient simultaneously to attain best effect. If deteriorated clinical conditions occurs, mechanical ventilation with extracorporeal circulation should be considered immediately. Surgical intervention must be initiated once adverse complications are found. With the utilization of various medical therapies, patients may survive fortunately but there is still a long period of time for his pulmonary rehabilitation.

Finally, immediate check of the PVL gene once sputum culture reveals the existence of *S. aureus* may help distinguish whether the isolate is likely the colonization or the true pathogen. Unnecessary antibiotics can also be avoided in combinations if PVL is negative; however, whether the routine genetic check of PVL in *S. aureus* is beneficial or not is still unknown. It needs more studies and large trials to be elucidated.

References

[1] Mehraj J, Witte W, Akmatov MK, Layer F, Werner G, Krause G. (2016) Epidemiology of *Staphylococcus aureus* nasal carriage patterns in the community. *Curr Top Microbiol Immunol.*, 398: 55-87.

[2] Wertheim HF, Melles DC, Vos MC, van Leeuwen W, van Belkum A, et al. (2005) The role of nasal carriage in *Staphylococcus aureus* infections. *Lancet Infect Dis.*, 5: 751–762.

[3] Cogen AL, Nizet V, Gallo R L. (2008) Skin microbiota: source of disease or defense. *Br J Dermatol.*, 158, 442–455.

[4] Kwiecinski JM, Horswill AR. (2020) *Staphylococcus aureus* bloodstream infections: pathogenesis and regulatory mechanisms. *Curr Opin Microbiol.*, 53, 51–60

[5] Josse J, Laurent F, Diot A. (2017) Staphylococcal adhesion and host cell invasion: Fibronectin-binding and other mechanisms. *Front Microbiol*, 8, 2433.

[6] Boyle-Vavra S, Daum RS. (2007) Community-acquired methicillin-resistant *Staphylococcus aureus*: the role of Panton–Valentine leucocidin. *Lab Invest.*, 87, 3-9.

[7] Gillet Y, Issartel B, Vanhems P, Fournet JC, Lina G, et al. (2002) Association between *Staphylococcus aureus* strains carrying gene for Panton-Valentine leukocidin and highly lethal necrotizing pneumonia in young immunocompetent patients. *Lancet*, 359, 753–759.

[8] Harkins CP, Pichon B, Doumith M, Parkhill J, Westh H, et al. (2017) MRSA emerged long before the introduction of methicillin into clinical practice. *Genome Biology.*, 18, 130.

[9] Bloomfield LE, Coombs GW, Tempone S, Armstrong PK. (2020) Marked increase in community-associated methicillin-resistant *Staphylococcus aureus* infections, Western Australia, 2004–2018. *Epidemiol Infect.*, 148, e153.

[10] Caitlyn RS, Fred CT, Richard VG. (2013) Transduction of staphylococcal cassette chromosome mec elements between strains of *Staphylococcus aureus*. *Antimicrob Agents Chemother.*, 57, 5233–5238.

[11] Coombs GW, Baines SL, Howden BP, Swenson KM, O'Brien FG. (2020) Diversity of bacteriophages encoding Panton-Valentine leukocidin in temporally and geographically related *Staphylococcus aureus*. *PLoS One*, 15, e0228676.

[12] Bhutia KO, Singh T, Adhikari L, Biswas S. (2015) Molecular characterization of community- & hospital-acquired methicillin-resistant & methicillin-sensitive *Staphylococcus aureus* isolates in Sikkim. *Indian J Med Res.*, 142, 330-335.

[13] Bhatta DR, Cavaco LM, Nath G, Kumar K, Gaur A, et al. (2016) Association of Panton Valentine Leukocidin (PVL) genes with methicillin resistant *Staphylococcus aureus* (MRSA) in Western Nepal: a matter of concern for community infections (a hospital based prospective study). *BMC Infect Dis.*, 16, 199.

[14] De Lencastre H, Oliveira D, Tomasz A. (2008) Antibiotic resistant *Staphylococcus aureus*: a paradigm of adaptive power. *Curr Opin Microbiol.*, 10, 428–435.

[15] Deurenberg RH, Stobberingh EE. (2008) The evolution of *Staphylococcus aureus*. *Infect Genet Evol.*, 8, 747–763.

[16] Chen CJ, Huang YC. (2014) New epidemiology of *Staphylococcus aureus* infection in Asia. *Clin Microbiol Infect.*, 20, 605-23.

[17] Klevens RM, Melissa AM, Joelle N, Petit S, Gershman K, et al. (2007) Invasive methicillin-resistant *Staphylococcus aureus* infections in the United States. *JAMA.*, 298, 1763-1771.

[18] Van Trijp MJ, Melles DC, Hendriks WD, Parlevliet GA, Gommans M, Ott A. (2007) Successful control of widespread methicillin-resistant *Staphylococcus aureus* colonization and infection in a large teaching hospital in the Netherlands. *Infect Control Hosp Epidemiol.*, 28, 970-975.

[19] Song JH, Hsueh PR, Chung DR, Ko KS, Kang CI, et al. (2011) Spread of methicillin-resistant *Staphylococcus aureus* between the community and the hospitals in Asian countries: an ANSORP study. *J Antimicrob Chemother.*, 66, 1061–1069.

[20] Kang CI, Song JH. (2013) Antimicrobial Resistance in Asia: Current epidemiology and clinical implications. *Infect Chemother*; 45, 22-31.

[21] Rasigade JP, Laurent F, Lina G, Meugnier H, Bes M, et al. (2010) Global distribution and evolution of Panton-Valentine leukocidin-positive methicillin-susceptible *Staphylococcus aureus*, 1981-2007. *J Infect Dis*., 201, 1589-1597.

[22] Darboe S, Dobreniecki S, Jarju S, Jallow M, Mohammed NI, et al. (2019) Prevalence of Panton-Valentine leukocidin (PVL) and antimicrobial resistance in community-acquired clinical *Staphylococcus aureus* in an urban Gambian hospital: A 11-year period retrospective pilot study. *Front Cell Infect Microbiol*., 9, 170.

[23] Brown ML, O'Hara FP, Close NM, Mera RM, Miller LA, et al. (2012) Prevalence and sequence variation of Panton-Valentine leukocidin in methicillin-resistant and methicillin-susceptible *Staphylococcus aureus* strains in the United States. *J Clin Microbiol*., 50, 86–90.

[24] Klein S, Hannesen J, Zanger P, Heeg K, Boutin S, Nurjadi D. (2020) Entry of Panton–Valentine leucocidin-positive methicillin-resistant *Staphylococcus aureus* into the hospital: prevalence and population structure in Heidelberg, Germany 2015–2018. *Sci Rep*., 10, 13243.

[25] Morrison MA, Hageman JC, Klevens RM. (2006) Case definition for community-associated methicillin-resistant *Staphylococcus aureus*. *J Hosp Infect*., 62, 241.

[26] Nemerovski CW, Klein KC. (2008) Community-associated methicillin-resistant *Staphylococcus aureus* in the pediatric population. *J Pediatr Pharmacol Ther*., 13, 212–225.

[27] Zetola N, Francis JS, Nuermberger EL, Bishai WR. (2005) Community-acquired meticillin-resistant *Staphylococcus aureus*: an emerging threat. *Lancet Infect Dis*., 5, 275-286.

[28] Tenover FC, McDougal LK, Goering RV, Killgore G, Projan SJ, et al. (2006) Characterization of a strain of community-associated methicillin-resistant *Staphylococcus aureus* widely disseminated in the United States. *J Clin Microbiol*., 44, 108-118.

[29] Abdullahi IN, Lozano C, Laura RR, Fernández-Fernández R, Zarazaga M, et al. (2021) Ecology and genetic lineages of nasal *Staphylococcus aureus* and MRSA carriage in healthy persons with or without animal-related occupational risks of colonization: A review of global reports. *Pathogens*, 10, 1000.

[30] Otto M. (2010) Staphylococcus colonization of the skin and antimicrobial peptides. *Expert Rev Dermatol*., 5, 183–195.

[31] Wu M, Tong X, Liu S, Wang D, Wang L, Fan H. (2019) Prevalence of methicillin-resistant *Staphylococcus aureus* in healthy Chinese population: A system review and meta-analysis. *PLoS One*., *14*, e0223599.

[32] Shallcross LJ, Fragaszy E, Johnson AM, Hayward AC. (2013) The role of the Panton-Valentine leucocidin toxin in staphylococcal disease: a systematic review and meta-analysis. *Lancet Infect Dis*., 13, 43–54.

[33] Holmes A, Ganner M, McGuane S, Pitt TL, Cookson BD, et al. (2005) *Staphylococcus aureus* isolates carrying Panton-Valentine leucocidin genes in England and Wales: Frequency, characterization, and association with clinical disease. *J Clin Microbiol*, 43, 2384–2390.

[34] Fernández S, Murzicato S, Sandoval O, Fernández-Canigia L, Mollerach M. (2015) Community-acquired necrotizing pneumonia caused by methicillin-resistant *Staphylococcus aureus* ST30-SCCmecIVc-spat019-PVL positive in San Antonio de Areco, Argentina. *Rev Argent Microbiol.* 47, 50-53.

[35] Honarpour N, Mao JT. (2007) A case of fatal community-acquired necrotizing pneumonia caused by Panton-Valentine leukocidin positive methicillin-sensitive *Staphylococcus aureus. Curr Resp Med Rev.*, 3, 49-51.

[36] Francis JS, Doherty MC, Lopatin U, Johnston CP, Sinha G, et al. (2005) Severe community-onset pneumonia in healthy adults caused by methicillin-resistant *Staphylococcus aureus* carrying the Panton-Valentine leukocidin genes. *Clin Infect Dis.*, 40, 100-107.

[37] Löffler B, Niemann S, Ehrhardt C, Horn D, Lanckohr C, et al. (2013) Pathogenesis of *Staphylococcus aureus* necrotizing pneumonia: the role of PVL and an influenza coinfection. *Expert Rev Anti Infect Ther.*, 11, 1041-1051.

[38] Chen J, Luo Y, Zhang S, Liang Z, Wang Y, et al. (2014) Community-acquired necrotizing pneumonia caused by methicillin-resistant *Staphylococcus aureus* producing Panton–Valentine leukocidin in a Chinese teenager: case report and literature review. *Int J Infect Dis.*, 26: 17–21.

[39] Elisabeth P, Maria S, Irene G, Helen G, Apostolos A. (2014) Success stories about severe pneumonia caused by Panton–Valentine leucocidin-producing *Staphylococcus aureus. Braz J Infect Dis.*, 18, 341–345.

[40] Yasir M, Khanchandani N, Lee S. (2013) Panton Valentine leukocidin (PVL) toxin positive methicillin-resistant *Staphylococcus aureus* (MRSA) necrotizing pneumonia associated with pseudomembranous necrotizing tracheobronchitis. *Expert Rev Anti Infect Ther.*, 11, 1041-51.

[41] Roberts JC, Gulino SP, Peak KK, Luna VA, Sanderson R. (2008) Fatal necrotizing pneumonia due to a Panton-Valentine leukocidin positive community-associated methicillin-sensitive *Staphylococcus aureus* and Influenza co-infection: a case report. *Ann Clin Microbiol Antimicrob.*, 7, 5.

[42] Larsen SAH, Kyhl K, Baig S, Petersen A, Av Steinum MR, et al. (2022) Life-threatening necrotizing pneumonia with Panton–Valentine leukocidin-producing, methicillin-sensitive *Staphylococcus aureus* in a healthy male co-Infected with influenza B. *Infect. Dis. Rep.*, 14, 12–19.

[43] Xia H, Gao J, Xiu M, Li D. (2020) Community-acquired pneumonia caused by methicillin-resistant *Staphylococcus aureus* in a Chinese adult: A case report. *Medicine.*, 99, e20914

[44] Duployez C, Guern RL, Tinez C, Lejeune AL, Robriquet L, et al. (2020). Panton-Valentine Leukocidin–secreting *Staphylococcus aureus* pneumonia complicating COVID-19. *Emerg Infect Dis.*, 26, 1939-1941.

[45] Nygaard U, Petersen A, Larsen AR, Rytter MJH, Hartling U, et al. (2022) Fatal SARS-CoV-2-associated Panton-Valentine leukocidin-producing Staphylococcal bacteremia. *Pediatr Infect Dis J.*, doi: 10.1097/INF.0000000000003476.

[46] Jung N, Lehmann C, Hellmann M, Seifert H, Valter MM, et al. (2008) Necrotizing pneumonia caused by Panton-Valentine leucocidin-producing *Staphylococcus*

aureus originating from a Bartholin's abscess. *Infect Dis Obstet Gynecol.*, 2008, 491401.

[47] Kuo PH, Chen FJ, Chan MC, Kan LP, Yeh KM, et al. (2016) Pyomyositis and necrotizing pneumonia in a young soldier caused by methicillin-susceptible *Staphylococcus aureus* of sequence type 59 lineage. *J Intern Med Taiwan*, 27, 49-54.

[48] Ahmad NI, Yean CY, Foo PC, Mohamad Safiee AW, Hassan SA. (2020) Prevalence and association of Panton-Valentine leukocidin gene with the risk of sepsis in patients infected with methicillin resistant *Staphylococcus aureus*. *J Infect Public Health.*, 13, 1508-1512.

[49] Lindsay JA, Holden MTG. (2004) *Staphylococcus aureus*: superbug, super genome? *Trends Microbiol.*, 12, 378-385.

[50] Malachowa N, DeLeo FR. (2010) Mobile genetic elements of *Staphylococcus aureus*. *Cell. Mol. Life Sci.*, 67, 3057–3071.

[51] Turner NA, Sharma-Kuinkel BK, Maskarinec SA, Eichenberger EM, Shah PP, et al. (2019) Methicillin-resistant *Staphylococcus aureus*: an overview of basic and clinical research. *Nat Rev Microbiol.*, 17, 203-218.

[52] Genestier AL, Michallet MC, Prévost G, Bellot G, Chalabreysse L, et al. (2005) *Staphylococcus aureus* Panton-Valentine leukocidin directly targets mitochondria and induces Bax-independent apoptosis of human neutrophils. *J Clin Invest.*, 115, 3117-3127.

[53] Lo WT, Wang CC. (2011) Panton-Valentine leukocidin in the pathogenesis of community-associated methicillin-resistant *Staphylococcus aureus* infection. *Pediatr Neonatol.*, 52, 59-65.

[54] Diep BA, Palazzolo-Ballance AM, Tattevin P, Basuino L, Braughton KR, et al. (2008) Contribution of Panton-Valentine leukocidin in community-associated methicillin-resistant *Staphylococcus aureus* pathogenesis. *PLoS One,* 3, e3198.

[55] Mazzoleni V, Zimmermann K, Smirnova A, Tarassov I, Prévost G. (2020) *Staphylococcus aureus* Panton-Valentine leukocidin triggers an alternative NETosis process targeting mitochondria. *FASEB J.*, 35, e21167.

[56] Holzinger D, Gieldon L, Mysore V, Nippe N, Taxman DJ, et al. (2012) *Staphylococcus aureus* Panton-Valentine leukocidin induces an inflammatory response in human phagocytes via the NLRP3 inflammasome. *J Leukoc Biol.*, 92, 1069–1081.

[57] Guillet V, Roblin P, Werner S, Coraiola M, Menestrina G, et al. (2004) Crystal structure of leucotoxin S component: new insight into the staphylococcal beta-barrel pore-forming toxins. *J Biol Chem.*, 279, 41028–41037.

[58] Pedelacq JD, Maveyraud L, Prevost G, Baba-Moussa L, González A, et al. (1999) The structure of a *Staphylococcus aureus* leucocidin component (LukF-PV) reveals the fold of the water-soluble species of a family of transmembrane pore-forming toxins. *Structure.*, 7, 277–287.

[59] Noda M, Kato I, Matsuda F, Hirayama T. (1981) Mode of action of staphylococcal leukocidin: relationship between binding of 125I-labeled S and F components of leukocidin to rabbit polymorphonuclear leukocytes and leukocidin activity. *Infect Immun.*, 34, 362–367.

[60] Yamamoto T, Nishiyama A, Takano T, Yabe S, Higuchi W, et al. (2010) Community-acquired methicillin-resistant Staphylococcus aureus: community transmission, pathogenesis, and drug resistance. *J Infect Chemother.*, 16, 225–254.

[61] Oogai Y, Matsuo M, Hashimoto M, Kato F, Sugai M, Komatsuzawa H. (2011) Expression of virulence factors by *Staphylococcus aureus* grown in serum. *Appl Environ Microbiol.*, 77, 8097–8105.

[62] Cruz AR, den Boer MA, Strasser J, Zwarthoff SA, Beurskens FJ, et al. (2021) Staphylococcal protein A inhibits complement activation by interfering with IgG hexamer formation. *Proc Natl Acad Sci USA.*, 118, e2016772118.

[63] Kobayashi SD, DeLeo FR. (2013) *Staphylococcus aureus* protein A promotes immune suppression. *mBio*, 4, e00764-13.

[64] Kaplan SL, Hulten KG, Mason EO. (2009) *Staphylococcus Aureus* Infections (Coagulase-positive Staphylococci). In: *Feigin and Cherry's Textbook of Pediatric Infectious Diseases*, Sixth Edition, Chapter 91, pages 1197-1213.

[65] Onogawa T. (2002) Staphylococcal alpha-toxin synergistically enhances inflammation caused by bacterial components. *FEMS Immunol Med Microbiol.*, 33, 15-21.

[66] Seeger W, Birkemeyer RG, Emert L, Suttorp N, Bhakdi S, Duncker HR. (1990) Staphylococcal alpha-toxin-induced vascular leakage in isolated perfused rabbit lungs. *Lab Invest.*, 63, 341–349.

[67] Morse SI. (1986) Specific microbial agents of disease. In: *Infectious Diseases and Medical Microbiology*, 2nd Edn. pp. 236–242.

[68] Murray PR, Drew WL, Kobayashi GS, and Thompson JH. (1990) Staphylococcus. In: *Medical Microbiology*, pp. 47–63.

[69] Niemann S, Ehrhardt C, Medina E, Warnking K, Tuchscherr L, et al. (2012) Combined action of influenza virus and *Staphylococcus aureus* panton-valentine leukocidin provokes severe lung epithelium damage. *J Infect Dis.*, 206, 1138–1148.

[70] Nathwani D, Morgan M, Masterton RG, Dryden M, Cookson BD, et al. (2008) Guidelines for UK practice for the diagnosis and management of methicillin-resistant *Staphylococcus aureus* (MRSA) infections presenting in the community. *J Antimicrob Chemother.*, 61, 976– 994.

[71] Planet PJ, Narechania A, Chen L, Mathema B, Boundy S, et al. (2017) Architecture of a species: Phylogenomics of *Staphylococcus aureus*. *Trends Microbiol.*, 25, 153-166.

[72] Dingle TC, MacCannell DR. (2015) Current and emerging technologies for the diagnosis of microbial infections. In: *Methods in Microbiology.* Chapter 9, pp. 329-458.

[73] Pérez-Losada M, Arenas M, Castro-Nallar E (2017) Multilocus sequence typing of pathogens: Methods, analyses, and applications. In: *Genetics and Evolution of Infectious Diseases,* 2nd Edn, pp. 383-404.

[74] Chen YJ, Chen PA, Chen CJ, Huang YC. (2019) Molecular characteristics and clinical features of pediatric methicillin-susceptible *Staphylococcus aureus* infection in a medical center in northern Taiwan. *BMC Infect Dis.*, 19, 402.

[75] Morgan MS. (2007) Diagnosis and treatment of Panton-Valentine leukocidin (PVL)-associated *Staphylococcal* pneumonia. *Int J Antimicrob Agents.*, 30, 289-296.

[76] Kravitz GR, Dries DJ, Peterson ML, Schlievert PM. (2005) Purpura fulminans due to *Staphylococcus aureus*. *Clin Infect Dis.*, 40, 941–947.

[77] Salliot C, Zeller V, Puechal X, Manceron V, Sire S, et al. (2006) Panton–Valentine leukocidin producing *Staphylococcus aureus* infections: report of 4 French cases. *Scand J Infect Dis.*, 38, 192–234.

[78] Torrell E, Molin D, Tabno E, Ehrenborg C, Ryden C. (2005) Community acquired pneumonia and bacteraemia in a healthy young woman caused by methicillin-resistant *Staphylococcus aureus* (MRSA) carrying the genes encoding Panton–Valentine leukocidin (PVL). *Scand J infect Dis.*, 7, 902–904.

[79] Bamberger DM. (2017) The role of Panton-Valentine leukocidin: The pendulum swings. *J Infect Dis*, 215, 1346–1348.

[80] Takigawa Y, Fujiwara K, Saito T, Nakasuka T, Ozeki T, et al. (2019) Rapidly progressive multiple cavity formation in necrotizing pneumonia caused by community-acquired methicillin-resistant *Staphylococcus aureus* positive for the Panton-Valentine leucocidin gene. *Intern Med.*, 58, 685-691.

[81] Pan F, Wang B, Zhang H, Shi Y, Xu Q. (2021) The clinical application of Filmarray respiratory panel in children especially with severe respiratory tract infections. *BMC Infect Dis.*, 21, 230.

[82] Ede S, Davis GM, Holmes FH. (1959) Staphylococci pneumonia. *JAMA.*, 170, 638–643.

[83] Hilton B, Tavare AN, Creer D. (2017) Necrotising pneumonia caused by non-PVL *Staphylococcus aureus* with 2-year follow-up. *BMJ Case Rep.*, 2017, bcr2017221779

[84] Banthia S, Meka VG, Pillai SK, Wener, K, Freedman, R, et al. (2005) A fatal case of necrotizing pneumonia caused by community-associated methicillin-resistant *Staphylococcus aureus*. *Infect Dis Clin Pract*; 13, 132–138.

[85] Gonzalez BE, Hulten KG, Dishop MK, Lamberth LB, Hammerman WA, et al. (2005) Pulmonary manifestations in children with invasive community-acquired *Staphylococcus aureus* infection. *Clin Infect Dis.*, 41, 583-590.

[86] Hoffer FA, Bloom DA, Colin AA, Fishman SJ. (1999) Lung abscess versus necrotizing pneumonia: implications for interventional therapy. *Pediatr Radiol.*, 29, 87–91.

[87] Haider S, Wright D. (2013) Panton-Valentine leukocidin *Staphylococcus* causing fatal necrotising pneumonia in a young boy. *BMJ Case Rep.*, 2013, bcr2012007655.

[88] Noah MA, Dawrant M, Faulkner GM, Hill AM, Harvey C, et al. (2010) Panton-valentine leukocidin expressing *Staphylococcus aureus* pneumonia managed with extracorporeal membrane oxygenation: Experience and outcome. *Crit Care Med.*, 38, 2250-2253.

[89] Zheng B, Jiang S, Xu Z, Xiao Y, Li L. (2015) Severe infective endocarditis with systemic embolism due to community associated methicillin-resistant *Staphylococcus aureus* ST630. *Braz J Infect Dis.*, 19, 85-89.

[90] Stevens DL, Ma Y, Salmi DB, McIndoo E, Wallace RJ, Bryant AE. (2007) Impact of antibiotics on expression of virulence associated exotoxin genes in methicillin-sensitive and methicillin-resistant *Staphylococcus aureus*. *J Infect Dis*., 19, 202-211.

[91] Dumitrescu O, Badiou C, Bes M, Reverdy ME, Vandenesch F, et al. (2008) Effect of antibiotics, alone and in combination, on Panton-Valentine leukocidin production by a *Staphylococcus aureus* reference strain. *Clin Microbiol Infect*., 14: 384–388.

[92] Claire ET, Shiranee S. (2015) Panton–Valentine leucocidin expression by *Staphylococcus aureus* exposed to common antibiotics. *J Infect*, 71: 338–346.

[93] Cataldo MA, Taglietti F, Petrosillo N. (2010) Methicillin-resistant *Staphylococcus aureus*: a community health threat. *Postgrad Med*, 122, 16-23.

[94] Dumitrescu O, Choudhury P, Boisset S, Badiou C, Bes M, et al. (2011) Beta-lactams interfering with PBP1 induce Panton-Valentine leukocidin expression by triggering sarA and rot global regulators of *Staphylococcus aureus*. *Antimicrob Agents Chemother*., 55, 3261–3271.

[95] Dumitrescu O, Boisset S, Badiou C, Bes M, Benito Y, et al. (2007) Effect of antibiotics on *Staphylococcus aureus* producing Panton-Valentine leukocidin. *Antimicrob Agents Chemother*., 51, 1515–1519.

[96] Bernardo K, Pakulat N, Fleer S, Schnaith A, Utermöhlen O, et al. (2004) Subinhibitory concentrations of linezolid reduce *Staphylococcus aureus* virulence factor expression. *Antimicrob Agents Chemother*., 48, 546-555.

[97] Lim WS, Baudouin SV, George RC, Hill AT, Jamieson C, et al. (2009) British Thoracic Society guidelines for the management of community acquired pneumonia in adults: update 2009. *Thorax*., 64, Suppl 3, iii 1-55.

[98] Nichani R, McGrath B. (2017) OSCEs for the final FFICM. *J Intensive Care Soc*; 18, 170.

[99] De Pascale G, Montini L, Pennisi M, Bernini V, Maviglia R, et al. (2014) High dose tigecycline in critically ill patients with severe infections due to multidrug-resistant bacteria. *Critical Care*, 18, R90.

[100] Karampela I, Poulakou G, Dimopoulos G. (2012) Community acquired methicillin resistant *Staphylococcus aureus* pneumonia: an update for the emergency and intensive care physician. *Minerva Anestesiol*., 78, 930-940.

[101] Gauduchon V, Cozon G, Vandenesch F, Genestier AL, Eyssade N, et al. (2004) Neutralization of *Staphylococcus aureus* Panton Valentine leukocidin by intravenous immunoglobulin *in vitro*. *J Infect Dis*., 189, 346-353.

[102] McGrath B, Rutledge F, Broadfield E. (2008) Necrotising pneumonia, *Staphylococcus aureus* and Panton-Valentine leukocidin. *J Intensive Care Soc*., 9, 170-172.

[103] Ek T, Andersson O, Kasemo AU, Wede M, Nilsson PA. (2007) [PVL positive *Staph aureus* as the cause of necrotizing pneumonia. Description of three severe cases in earlier healthy young persons]. *Lakartidningen*, 7, 509–513. [Swedish]

[104] Mosnier LO, Gale AJ, Yegneswaran S, Griffin JH. (2004) Activated protein C variants with normal cytoprotective but reduced anticoagulant activity. *Blood*., 104, 1740-1744.

[105] Griffin JH, Fernández JA, Gale AJ, Mosnier LO. (2007) Activated protein C. *J Thromb Haemost*., 5, Suppl 1, S73-S80.

[106] Fujisaki N, Takahashi A, Arima T, Mizushima T, Ikeda K, et al. (2014) Successful treatment of Panton-Valentine leukocidin-expressing *Staphylococcus aureus*-associated pneumonia co-infected with influenza using extracorporeal membrane oxygenation. *In Vivo.*, 28, 961-966.

[107] Lavoue S, Gac GL, Gacouin A, Revest M, Sohier L, et al. (2016) Extracorporeal circuit for Panton-Valentine leukocidin-producing *Staphylococcus aureus* necrotizing pneumonia. *Med Mal Infect.*, 46, 314–317.

[108] Cuddihy J, Patel S, Mugha N l, Lockie C, Trimlett R, et al. (2021) Near-fatal Panton-Valentine leukocidin-positive *Staphylococcus aureus* pneumonia, shock and complicated extracorporeal membrane oxygenation cannulation: A case report. *World J Crit Care Med.*, 10, 301-309.

[109] MacLaren G, Butt W. (2007) Extracorporeal membrane oxygenation and sepsis. *Crit Care Resusc.*, 9, 76–80.

[110] Lees NJ, Rosenberg AJP, Hurtado-Doce AI, Jones J, Marczin N, et al. (2016) Combination of ECMO and cytokine adsorption therapy for severe sepsis with cardiogenic shock and ARDS due to Panton–Valentine leukocidin—positive *Staphylococcus aureus* pneumonia and H1N1. *J Artif Organs.*, 19, 399-402.

[111] Poli EC, Simoni C, André P, Buclin T, Longchamp D, et al. (2019) Clindamycin clearance during Cytosorb® hemoadsorption: A case report and pharmacokinetic study. *Int J Artif Organs.*, 42, 258-262.

[112] Song M, Winchester J, Albright RL, Capponi VJ, Choquette MD, Kellum JA. (2004) Cytokine removal with a novel adsorbent polymer. *Blood Purif.*, 22, 428–434.

[113] Morris C, Gray L, Giovannelli M. (2015) Early report: the use of Cytosorb haemo absorption column as an adjunct in managing severe sepsis: initial experiences, review and recommendations. *J Intensive Care Soc.*, 16, 257–264.

[114] Hetz H, Berger R, Recknagel P, Steltzer H. (2014) Septic shock secondary to beta-hemolytic streptococcus-induced necrotizing fasciitis treated with a novel cytokine adsorption therapy. *Int J Artif Organs.*, 37, 422–426.

[115] Basu R, Pathak S, Goyal J, Chaudhry R, Goel RB, et al. (2014) Use of a novel hemadsorption device for cytokine removal as adjuvant therapy in a patient with septic shock with multi-organ dysfunction: a case study. *Indian J Crit Care Med*, 18, 822–824.

[116] Bruenger F, Kizner L, Weile J, Morshuis M, Gummert JF. (2015) First successful combination of ECMO with cytokine removal therapy in cardiogenic septic shock: a case report. *Int J Artif Organs.*, 38, 113–116.

[117] Hawchar F, László I, Öveges N, Trásy D, Ondrik Z, Molnar Z. (2019) Extracorporeal cytokine adsorption in septic shock: A proof of concept randomized, controlled pilot study. *J Crit Care.*, 49, 172-178.

[118] Chow SH, Deo P, Yeung ATY, Kostoulias XP, Jeon Y, et al. (2020) Targeting NLRP3 and Staphylococcal pore-forming toxin receptors in human-induced pluripotent stem cell-derived macrophages. *J Leukoc Biol.*, 108, 967-981.

[119] Lopez-Aguilar C, Perez-Roth E, Mendez-Alvarez S, Duran MC, Casanova C, et al. (2007) Association between the presence of the Panton-Valentine leukocidin-

encoding gene and a lower rate of survival among hospitalized pulmonary patients with staphylococcal disease. *J Clin Microbiol.*, 45, 274–276.

[120] Rubinstein E, Kollef MH, Nathwani D. (2008) Pneumonia caused by methicillin-resistant *Staphylococcus aureus*. *Clin Infect Dis.*, 46, Suppl 5, S378–S385.

[121] Tong A, Tong SYC, Zhang Y, Lamlertthon S, Sharma-Kuinkel BK, et al. (2012) Panton-Valentine leukocidin is not the primary determinant of outcome for *Staphylococcus aureus* skin Infections: Evaluation from the CANVAS studies. *PLoS One*, 7, e37212.

[122] Lina G, Piémont Y, Godail-Gamot F, Bes M, Peter MO, et al. (1999) Involvement of Panton-Valentine leukocidin-producing *Staphylococcus aureus* in primary skin infections and pneumonia. *Clin Infect Dis.*, 29, 1128-1132.

[123] Li HT, Zhang TT, Huang J, Zhou YQ, Zhu JX, Wu BQ. (2011) Factors associated with the outcome of life-threatening necrotizing pneumonia due to community-acquired *Staphylococcus aureus* in adult and adolescent patients. *Respiration.*, 81, 448-460.

[124] Sicot N, Khanafer N, Meyssonnier V, Dumitrescu O, Tristan A, et al. (2013) Methicillin resistance is not a predictor of severity in community-acquired *Staphylococcus aureus* necrotizing pneumonia—results of a prospective observational study. *Clin Microbiol Infect.*, 19, E142-148.

[125] Vardakas KZ, Matthaiou DK, Falagas ME. (2009) Comparison of community-acquired pneumonia due to methicillin-resistant and methicillin-susceptible *Staphylococcus aureus* producing the Panton-Valentine leukocidin. *Int J Tuberc Lung Dis.*, 13, 1476–1485.

Biographical Sketch

Wen-Liang Yu

Affiliation: Chi-Mei Medical Center, Tainan City, Taiwan

Education:

- University: School of Chinese Medicine, College of Chinese Medicine, China Medical University, Taichung City, Taiwan, 1980 September ~1987 June
- Clerk & Internship: Cathay General Hospital, 1985 July~1987 June
- Residence: Department of Internal Medicine, China Medical University Hospital, Taichung, Taiwan, 1989 July-1994 June

Research and Professional Experience:

- Research fellow: Division of Infectious Diseases, Department of Internal Medicine, Taipei Veteran General Hospital, Taipei, Taiwan, 1993-1994
- Research fellow: Divisions of Microbiology and Medical Research, the University of Iowa, Iowa, USA, 2000-2001
- Attending Physician in Division of Infectious Diseases, Department of Internal Medicine, China Medical University Hospital, Taichung, Taiwan, 1994 July-2003 June
- Assistant Professor in Department of Internal Medicine, China Medical University, Taichung, Taiwan, 2001 July-2003 June.
- Assistant Professor in Department of Internal Medicine, Taipei Medical University, Taipei, Taiwan, 2004-2006

Current Professional Appointments:

- Attending Physician in Department of Intensive Care Medicine, Chi-Mei Medical Center, Tainan, Taiwan, 2003 July-now
- Associate Professor in Department of Internal Medicine, Taipei Medical University, Taipei, Taiwan, 2006-now

Honors:

1. Outstanding contributor in reviewing, International Journal of Infectious Diseases, in 2014,
2. Outstanding contributor in reviewing, International Journal of Infectious Diseases, in 2015.
3. Outstanding contributor in reviewing, Journal of Microbiology, Immunology, and Infection, 2015
4. Outstanding oral report (Second place) in Annual meeting of Society of Emergency & Critical Care Medicine, Taiwan, 2016
5. Award for Excellent Performance of Poster in Annual meeting of Society of Emergency & Critical Care Medicine, Taiwan, 2016
6. Guest Editor for a special issue in the Journal of Fungi, 2022

Biography:

Wen-Liang Yu is a physician at the Department of Intensive Medicine of Chi Mei Medical Center at Tainan City and is an associate professor of Department of Medicine, School of Medicine, College of Medicine of Taipei Medical

University at Taipei City in Taiwan. He's an expert in the infectious diseases especially in critically ill patients. Dr. Yu is also the author of 3 chapters in UpToDate, a well-known and well-marketed software system as an evidence-based clinical resource, and he has contributed to 7 chapters published by Nova Science Publishers in New York, 8 chapters published in SM Group eBook system, and over 100 papers published in SCI journals.

Publications: linked to Website:
https://tmu.pure.elsevier.com/en/persons/wen-liang-yu-2/publications/

Chapter 5

Biochemical Thermodynamics

E. Méndez*, PhD

Biomaterials Laboratory, Institute of Biological Chemistry, Faculty of Sciences, University of the Republic, Montevideo, Uruguay

Abstract

Since the studies pioneered by Hans Krebs in the '50s, the thermodynamic basis of enzyme-catalyzed reactions has received considerable attention. After the implementation of the biochemical reference state, the prediction power of thermodynamic calculations has increased notably. However, there is still a gap in the amount of information that prevents the wide application of biochemical thermodynamics to different processes. Group-contribution method, quantum chemical calculations, and big data science emerge as a solution to this problem. In this chapter, biochemical thermodynamics is considered after a brief historical revision, some basic thermodynamic foundations, the introduction of the different reference states, as well as the curated source of thermodynamic and experimental data that allow for the precise calculation of missing data. Novel calculation procedures will be considered. Finally, the reunification of chemical and biochemical thermodynamics will be discussed.

Keywords: balanced biochemical equations, group-contribution method, quantum chemical calculations, thermodynamic properties, transformed thermodynamic properties

---

* Corresponding Author's Email: emendez@fcien.edu.uy.

In: Advances in Biology. Volume 1
Editor: Charles D. Grant
ISBN: 979-8-88697-201-6

One Hundred Years of Biochemical Thermodynamics

The study of the aspartate-ammonia lyase reaction was the first reported measurement of the equilibrium constant of an enzyme-catalyzed reaction (Quastel and Woolf 1926). The authors did the measurements in the presence of bacteria containing the enzymatic system and provided unequivocal evidence of the catalytic action. At the same time, the determination of the equilibrium constant for the analyzed reaction suggested the possible use of these systems linked to the thermodynamic theory. This kind of study, though continued in the '30s and the '40s (Alberty 2004), received a considerable impulse in the '50s, with the studies performed by Kenneth Burton and Hans Krebs. In particular, the studies of biochemical reactions involving the hydrolysis of phosphate group-containing metabolites (Burton and Krebs 1953) followed the idea presented by Lipman several years before on the concept of "energy-rich phosphate bond" for the phosphate anhydride linkage in ATP molecule (Lipmann 1941). These initial works constitute key milestones in the evolution of biochemical thermodynamics, which history continues developing up until now.

The studies performed in those years led to a comprehensive book on biochemical thermodynamics that included an Appendix containing the first published Table of Biochemical Thermodynamic Data, arranged and curated by Kenneth Burton (Krebs et al. 1957), defining a new reference state: the biological reference state, which considers a fixed pH of 7. More complete tables were presented later (Thauer, Jungermann, and Decker 1977, Miller and Smith-Magowan 1990), and comprehensively arranged in the book "Thermodynamic Data for Biochemistry and Biotechnology" (Hinz and Hans-Jurgen 1986).

In the '80s and '90s, several studies on the thermodynamic properties and equilibrium measurements of enzyme-catalyzed reactions were led by Robert N. Goldberg and co-workers (Tewari, Steckler, and Goldberg 1988, Tewari et al. 1988, Goldberg and Tewari 1989, Goldberg, Tewari, and Ahluwalia 1989, Tewari and Goldberg 1989, Alberty and Goldberg 1993, A. Alberty and N.Goldberg 1994), who generated a database big enough to set the experimental foundation of biochemical thermodynamics.

Prof. Robert A. Alberty (1921 – 2014), from the Massachusetts Institute of Technology, was a key researcher in the foundation of biochemical thermodynamics. Details on his work were given by himself (Alberty 2010) in an issue dedicated to his contribution to the area by the *Journal of Physical Chemistry* (Cornish-Bowden, Goldberg, and Hammes 2010). Based mainly on the work and definitions given by Alberty, in 1994, the International Union of Biochemistry and Molecular Biology (IUBMB) set the foundations of Biochemical Thermodynamics, providing new definitions (transformed properties, biochemical reactions) and rules (nomenclature). In addition, a new reference state was defined: the biochemical reference state, which includes a fixed pH = 7 and pMg = 3, superseding the biological reference state (Alberty 1994b). In 2011, the rules for biochemical thermodynamics were updated, and several suggestions were given to researchers to provide experimental data within homogenized criteria (Alberty et al. 2011).

The differentiation between chemical and biochemical thermodynamics has been questioned recently, and a new commission joined by the International Union of Pure and Applied Chemistry (IUPAC) and IUBMB began the study on the possible reunification (Sabatini et al. 2021).

Thermodynamic Foundations

The thermodynamic theory is based on two (axiomatic) laws that are supported by extensive experimental evidence. Such laws deal with the conservation of energy (First Law) and the natural course of a process (Second Law). To mathematically express such laws, two kinds of quantities are considered: one group that depends on the equilibrium states of the system before and after the process, and another that depends on the path taken to perform the process (Engel and Reid 2020). The first ones are named *state functions* or *properties* and, for a given process, their changes are expressed by an exact differential, in such a way that the *net change* only depends on the equilibrium values before (Y_i) and after (Y_f) the process.

$$\Delta Y = \int_{Y_i}^{Y_f} dY = Y_f - Y_i \qquad (1)$$

The second kind of quantities are path-dependent and as a consequence, the *net amount* can only be calculated with the exact knowledge of the process. For such quantities, the value involved cannot be expressed by an exact differential, and the net amount is calculated by a line integral along the "path" employed for the process:

$$Y = \int_{path} \delta Y \qquad (2)$$

where the symbol "δ" denotes the inexact differential.

The First Law of Thermodynamics introduces the internal energy (*U*) of a system, stating that for a given process, its change (*dU*) is expressed by:

$$dU = \delta q + \delta w \tag{3}$$

where δq and δw represent the infinitesimal amount of heat flow into the system and work done on the system, respectively. Notice that the latter quantities are expressed as inexact differentials, as their net amount depends on the path taken to perform the process. As a direct consequence of the First Law, the enthalpy (*H*) is defined as a combination of thermodynamic properties:

$$H \equiv U + PV \tag{4}$$

where *P* and *V* are the pressure and the volume of the system.

The Second Law of Thermodynamics defines a new property, the entropy (*S*), whose infinitesimal change for a given process is given by:

$$dS = \frac{\delta q_{rev}}{T} \tag{5}$$

where *T* is the temperature. Again, notice that the infinitesimal heat flow to the system (δq_{rev}) cannot be expressed by an exact differential.

The combination of the First and Second laws of Thermodynamics introduces a new function that can also be expressed through an exact differential, the Gibbs energy:

$$G \equiv H - TS \tag{6}$$

The change in Gibbs energy due to a given process is related to the spontaneity of the process: when it is negative, the process is spontaneous.

Why Exact Differentials?

The advantage of using state functions (exact differentials) is that they are additive properties. As a consequence, a given path can be subdivided into several real or imaginary *N* steps characterized by a partial state function ($\Delta_{step}Y$), so the net value for the change in the property for the whole process (*pr*) is:

$$\Delta_{pr}Y = \sum_{step=1}^{step=N} \Delta_{step}Y \tag{7}$$

As we will see later, this condition forms the basis of the group-contribution methods for the calculation of thermodynamic properties. Likewise, biochemical compounds are found in different states depending on the pH and the presence of Mg^{2+} ions; for example, ATP can be in the form of H_4ATP, H_3ATP^-, H_2ATP^{2-}, $HATP^{3-}$, $MgATP^{2-}$, or $MgHATP^-$. Each of these forms is called a pseudoisomer, and the proportion in which each of them is found depending on given conditions of pH and pMg determines the final value for the thermodynamic properties of ATP (Alberty 1995).

Interpretation of the Thermodynamic Properties

The exact mechanisms of the work of metabolic systems are not always known, but it is possible to know the initial and final states related to a certain biological process. On that note, the thermodynamic properties, particularly the state functions, are the appropriate descriptors, since they only depend on the initial and final states of a process. The analysis of Gibbs energies, enthalpies, and entropies can be done beyond knowing the exact mechanism of the biochemical process while providing important information for understanding the consequences of the process under study. The determination of the thermodynamic properties of biochemical reactions constitutes one of the most powerful tools for understanding how biological systems work, including, in particular, their metabolism.

According to the Second Law of Thermodynamics, the spontaneity condition of a process is given by the increase in the entropy of the universe, $\Delta_{univ}S > 0$:

$$\Delta_{univ}S = \Delta_{pr}S + \Delta_{env}S \tag{8}$$

The combination of the First and Second laws of Thermodynamics provides the criterion for spontaneity seen from the system undergoing the process, $\Delta_{pr}G < 0$ (exergonic process):

$$\Delta_{pr}G = \Delta_{pr}H - T\Delta_{pr}S \tag{9}$$

The sign of the thermodynamic properties in a given process is associated with some very important characteristics: $\Delta_{pr}H > 0$ is associated with reactions that absorb heat at constant pressure (endothermic) and those that release heat have $\Delta_{pr}H < 0$ (exothermic). The environment (usually an aqueous solution) is the source or sink of the energy exchanged as heat, and in homeothermic animals, this heat is directly related to the entropy change in the environment, $\Delta_{env}S$. The process with $\Delta_{pr}H < 0$ delivers heat to the environment, producing an entropy increase, $\Delta_{env}S > 0$. Thus, it is feasible that some processes associated with $\Delta_{pr}S < 0$, such as biological self-assembly, are spontaneous. Conversely, endothermic processes with $\Delta_{pr}S < 0$ are not possible under any experimental conditions ($\Delta_{pr}G > 0$, endergonic process).

The Reference States

Advances in biochemical thermodynamics studies went hand in hand with the recognition of the importance of the physicochemical environment in which the process is carried out (Méndez and Cerdá 2016). In chronological order, the need to modify the chemical reference state (CRS) defines what is currently known as the biological reference state (BRS), which considers a fixed pH value of 7. Subsequently, the recognition of the incidence of Mg^{2+} ions in biochemical reactions, by forming complexes with many important metabolites, as well as the multiplicity of chemical equilibria between the intervening species, led to the definition of the biochemical reference state (BqRS), for which a restriction is added for the concentration of Mg^{2+} ions, pMg = 3. Temperature and ionic strength were maintained constant at 25°C and I = 0 M, respectively. Finally, the consideration of an ionic strength of 0.25 M and a temperature of 37°C define the near-physiological reference state (NPRS) (Table 1).

Table 1. Reference states for thermodynamic data (Méndez and Cerdá 2016)

Reference state	Constrains	Nomenclature[a]
Chemical (CRS)	1 bar, 25°C	$\Delta_{process}X_m^o$
Biological (BRS)	1 bar, 25°C, pH = 7	$\Delta_{process}X_m^{o\prime}$
Biochemical (BqRS)	1 bar, 25°C, pH = 7, pMg = 3, I = 0 M	$\Delta_{process}X_m^{\prime o}$
Near-physiological (NPRS)	1 bar, 37°C, pH = 7, pMg = 3, I = 0.25 M	$\Delta_{process}X_m^{\prime o}$

[a] X represents any thermodynamic property (H, G, S, C_p).

Calculations of thermodynamic properties at different reference states for the same reaction give rise to different results, and consequently, the biochemical relevance is different. For example, the pyrophosphate hydrolysis reaction turns out to be an endergonic process in the CRS, and an exergonic process in the BqRS (Mendez 2020).

Correction Terms for $\Delta_r G_m^o$ due to Temperature, pH, and Ionic Strength

Once the thermodynamic properties have been calculated in the CRS, some adjustments must be introduced to transform these values to the other reference states due to the incidence of pH, ionic strength, and temperature (Li et al. 2011). The calculation of the incidence of pH is simple to perform, and only affects those reactions that exchange H^+. The concentration of H^+ in the BRS is 10^{-7} M (pH = 7). Hence, for a given reaction at a temperature *T*, the correction term Δ_{pH} is:

$$\Delta_{pH} = -2.303\vartheta_H RTpH \tag{10}$$

where ϑ_H represents the stoichiometric number for H^+ *produced* in the reaction.

For the incidence of the ionic strength, the correction term Δ_I takes into account the extended Debye-Hückel equation (Alberty 2007):

$$\Delta_I = -RT\alpha_T\left(\frac{\sqrt{I}}{1+B\sqrt{I}}\right)\sum_i \vartheta_i z_i^2 \tag{11}$$

where z_i is the charge of the species *i*, ϑ_i is its stoichiometric coefficient, *B* = 1.6 $M^{-1/2}$ is an empirical constant, and α_T is the temperature-dependent coefficient in the Debye equation:

$$\alpha_T = 1.10708 - (1.54508\text{x}10^{-3})T + (5.95584\text{x}10^{-6})T^2 \tag{12}$$

The corrective term for temperature Δ_T arises from the van't Hoff equation, assuming that $\Delta_r H_m^o$ is independent of the temperature in the considered range:

$$\Delta_T = \Delta_r G^o_{298.15K}\left(\frac{T}{298.15} - 1\right) + \Delta_r H^o_{298.15K}\left(1 - \frac{T}{298.15}\right) \tag{13}$$

Finally, the variation of the Gibbs energy at any temperature and ionic strength can be derived from the tabulated thermodynamic properties in the CRS, based on the fact that the Gibbs energy is an additive property:

$$\Delta_r G^o_{T,I} = \Delta_r G^o_{298.15K,I=0} + \Delta_{pH} + \Delta_I + \Delta_T \tag{14}$$

Some examples of calculations to correct the Gibbs energies as a function of the experimental conditions can be consulted in the literature (Jin et al. 2020, Xie, Dai, and Hu 2016, Xie et al. 2014, Zhu et al. 2021).

Experimental Sources of Thermodynamic Data

In the study of enzyme-catalyzed reactions, it is usually necessary to use buffers and co-factors that allow the optimal condition for an enzyme to function. Likewise, these experimental conditions may not be homogeneous between the different studies, and therefore the experimental data are often difficult to compare. This fact was considered (Alberty et al. 2011) and led to the research reporting guidelines summarized in Table 2.

Table 2. Summary of recommendations for reporting enzyme-catalyzed thermodynamic data (Goldberg 2014)

Focus	Recommendation
The enzymatic system under study	• Reaction and stoichiometry specified • Aggregation state of reactants and products specified • Balanced chemical or biochemical equation according to specific rules (IUPAC and IUBMB, respectively)
Experimental procedures	• Purity of chemicals employed • Proper identification of chemicals through IUPAC names and/or CAS registry numbers • Fully detailed methodologies
Data report	• Mandatory report of all numerical values in the correct units (graphs may be also used but as a complement of numerical values) • Clear statement of the reference state employed (chemical, biological, biochemical, or near-physiological)

The main direct experimental source of reaction enthalpies are calorimetric measurements, while apparent equilibrium constants (*K*) are the source of Gibbs energies. Equilibrium constant measurements are usually simpler than calorimetric measurements, so there are more $\Delta_f G_m^o$ values reported than $\Delta_f H_m^o$ values. On the other hand, the possibility of experimentally measuring equilibrium constants is limited to a certain range of values, since very high or very low values of equilibrium constants require powerful analytical methodologies (Fuchs and Gessner 2001). In general, $K > 10^6$ or $K < 10^{-6}$ requires methodologies capable of detecting concentrations of the μM order. New analytical developments progressively push these limits, and the number of methods to experimentally determined *K* is increasing. For example, liquid chromatography-mass spectrometry (LC-MS) allows the simultaneous identification and quantification of metabolites, and high-resolution mass spectrometry and multi-isotope tracing are currently used as quantitative techniques under physiological conditions (Xu et al. 2020).

Calorimetry

Under conditions of constant pressure, the heat transferred during a biochemical reaction is equal to $\Delta_r H$. In this way, it is only necessary to know the equilibrium properties before and after the process to know the exchanged heat, without knowing exactly the process's trajectory. If the reaction is carried out in an adiabatic vessel, the heat due to the reaction is only exchanged with the environment (the aqueous solution), and only the temperature before and after the reaction must be measured:

$$Q = \Delta_r H = -\Delta_{env} H = C_p \Delta T \quad (15)$$

Equilibrium Data

Thermodynamic measurements made on systems in equilibrium can be used to calculate the missing thermodynamic properties of some compounds. Although these measurements are directly related to Gibbs energy, they are not the values that are usually tabulated, although they are part of the values that are carefully curated by specialists.

The determination of the equilibrium constants provides data in two ways: to calculate values of Gibbs energy of a reaction:

$$\Delta_r G^o = -RTln\,K \tag{16}$$

or to determine the values of enthalpy of reaction at a given temperature range (usually at 25°C and 37°C), through the van't Hoff equation:

$$\left[\frac{\partial \ln(K)}{\partial(1/T)}\right]_P = -\frac{\Delta_r H^o}{R} \tag{17}$$

Combustion Enthalpies

Combustion enthalpies are determined in bomb calorimeters, which are diathermic rigid-walled containers in which complete combustion of an organic compound takes place. The resulting heat of reaction corresponds to the change in the internal energy of the system $\Delta_c U$, which is related to the enthalpy of combustion $\Delta_c H$ according to:

$$\Delta_c H = \Delta_c U + RT\Delta n \tag{18}$$

where Δn is the net change in the gaseous moles during the reaction.

Curation of Experimental Data

The main sources of experimental data are the specialized journals, which function as a repository of data, with a quality ensured by the peer review system. Some of these journals are:

- *Journal of Physical and Chemical Reference Data*. A peer-reviewed scientific journal published by AIP Publishing with the editorial expertise of the National Institute of Standards and Technology (NIST), which provides critically evaluated physical and chemical property data, fully documented sources, and the criteria used for evaluation, including uncertainty analyses (Acree and Chickos 2021). Data in this Journal are mostly referred to as the chemical standard state.

- *The Journal of Chemical & Engineering Data.* A peer-reviewed scientific journal from the Americal Chemical Society (ACS), publishes data on the phase behavior and the physical, thermodynamic, and transport properties of well-defined materials, including biological samples.
- *The Journal of Chemical Thermodynamics.* A peer-reviewed scientific journal from Elsevier, whose objective is the dissemination of significant new knowledge in experimental equilibrium thermodynamics and transport properties of chemical systems.

Due to the inconsistencies in the experimental procedures and conditions, the curation of experimental data turns into a key step in their utilization for thermodynamic calculations (Méndez and Cerdá 2016). For biochemical data, the work of Dr. Robert N. Goldberg from NIST in the curation of such data has been monumental. Data were carefully arranged according to the enzymes involved: oxidoreductases (Goldberg et al. 1993), transferases (Goldberg and Tewari 1994b), hydrolases (Goldberg and Tewari 1994a), lyases (Goldberg and Tewari 1995a), and isomerases (Goldberg and Tewari 1995b), followed by two updates involving different types of enzymes (Goldberg, Tewari, and Bhat 2007, Goldberg 1999).

Calculation of the Thermodynamic Properties

The fundamental data for carrying out thermodynamic calculations are the thermodynamic properties of the formation of compounds, to which are added the standard molar heat capacity data, all of them in a clearly defined standard reference state. Therefore, having tables that summarize these data, coming from reliable sources and having been curated by experts, constitutes one of the most powerful tools that thermodynamics has.

Based on the thermodynamic data obtained experimentally, it is possible to calculate missing enthalpies or Gibbs energies of formation of different compounds for a given reference state. For a general reaction:

$$Reactants \rightarrow Products$$

by the additive properties of the state functions, it is possible to relate the thermodynamic properties of reaction to those of formation:

$$\Delta_r X = \sum_i \vartheta_i \Delta_f X_i \quad (19)$$

where ϑ_i represents the stoichiometric coefficient (positive for products, negative for reactants), $\Delta_f X_i$ are the formation properties of each species, and the quantity $\Delta_r X$ can refer to any reaction: formation, combustion, ionization, etc.

From Formation Reactions Data

The formation properties of a given compound are taken as references for thermodynamic calculations. A chemical reaction that forms one mole of product from their component pure elements under chemical standard reference conditions (1 bar and 298.15 K) is called a *formation reaction in the standard state*. All thermodynamic properties related to this reaction are known as *standard properties of formation*. By definition, the formation Gibbs energy and enthalpy of the pure elements in their most stable state of aggregation are assigned a value of 0 in the standard reference state (Table 3).

Table 3. The physical state and thermodynamic properties of key elements in the chemical reference state conditions of 1 bar pressure, T = 298.15 K, I = 0 M (Haynes 2016)

Atom	**Compound**	$\Delta_f H_m^o$ **(kJ mol^{-1})**	$\Delta_f G_m^o$ **(kJ mol^{-1})**	S_m^o **(J mol^{-1} K^{-1})**
C	C (graphite)	0	0	5.74
C	CO_2 (g)	-393.509	-394.359	213.74
H	H_2 (g)	0	0	130.680
O	O_2 (g)	0	0	205.152
O	H_2O (l)	-285.830	-237.141	69.95
N	N_2 (g)	0	0	191.609
P	P (white)	0	0	41.09
S	S (rhombic)	0	0	32.054
S	SO_2 (g)	-296.830	-300.194	248.22

For a general compound $H_mC_nO_pN_qS_rP_t$, the formation reaction is formulated as:

$$\frac{m}{2}H_{2\,(g)} + n\,C_{(gr)} + \frac{p}{2}O_{2\,(g)} + \frac{q}{2}N_{2\,(g)} + r\,S_{(rh)} + t\,P_{(wh)} \rightarrow H_mC_nO_pN_qS_rP_t$$

The formation entropy has a finite value as the null value for this property corresponds to 0 K (Table 3). As most of the data for compounds refers to $\Delta_f H_m^o$ and $\Delta_f G_m^o$, an approach to absolute entropy data comes from the calculation of the entropy of the formation reaction:

$$\Delta_f S_m^o = \frac{\Delta_f H_m^o - \Delta_f G_m^o}{T} \quad (20)$$

from which S_m^o of the desired compound can be calculated (Alberty 2006b):

$$S_m^o(compound) = \Delta_f S_m^o(compound) + \sum_{element} N_{element}\, S_m^o(element) \quad (21)$$

where $N_{element}$ is the number of atoms of a given element in the compound.

Notice that this calculation yields the absolute entropy of the compound but in the standard reference state, which may differ from its value in aqueous solutions. In this case, the formation reaction of a charged species can be written as:

$$\frac{m}{2}H_{2\,(g)} + nC_{(gr)} + \frac{p}{2}O_{2\,(g)} + \frac{q}{2}N_{2\,(g)} + rS_{(rh)} + tP_{(wh)} \rightarrow \left[C_nO_pN_qS_r\right]^{m-}(aq) + mH^+(aq)$$

According to Alberty (Alberty 2005), there is a convention problem for H^+ ions in thermodynamic tables. Note that if $\Delta_f H_m^o$ (H^+) = 0 and $\Delta_f G_m^o$ (H^+) = 0 at $I = 0$ (at $I \neq 0$ should be corrected according to eq. (11)), then the formation properties of the solid compound would be the same as its ionic form. If we consider the formation reaction of H^+ as:

$$\frac{1}{2}H_2\,(g) = H^+\,(aq) + e^-$$

where e^- is the formal electron. If we assign a value of 0 to the formation properties of the formal electron, then we can write:

$$\Delta_f S^o\,[H^+(aq)] = S_m^o\,[H^+(aq)] + S_m^o\,[e^-] - \frac{1}{2}S_m^o\,[H_2(g)] = 0 \quad (22)$$

and consequently, $S^o_m\ [H^+(aq)] = \frac{1}{2} S^o_m\ [H_2(g)]$, rather than zero.

From Combustion Reactions Data

The complete combustion reaction for a general compound $H_mC_nO_pN_qS_r$ is:

$$H_mC_nO_pN_qS_r\ (s) + \frac{(4n + m - 2p + 4r)}{4}\ O_2\ (g) \rightarrow nCO_2\ (g) + \frac{m}{2}\ H_2O\ (l) + \frac{q}{2}\ N_2\ (g) + rSO_2\ (g)$$

and knowing the values for the enthalpies of formation of the compounds involved (Table 3), the enthalpy of formation of the biochemical compound can be calculated. Some examples in the literature illustrate these calculations (Boerio-Goates et al. 2001, Emel'yanenko, Zaitsau, and Verevkin 2017).

From Apparent Equilibrium Data

The measurements of the apparent equilibrium constants at the standard pressure (1 bar) and different temperatures allow for the estimation of the enthalpy of the reaction ($\Delta_r H^o$) under study. This estimate is based on the assumption that $\Delta_r H^o$ is independent of temperature over the working temperature range; in this way, from the van't Hoff equation (eq. 17), $\Delta_r H^o$ can be calculated from the experimental slope of the graph ln(*K*) vs. (1/*T*). The difference between the standard temperature (298.15 K) and the physiological temperature (310.15 K) is low enough to ensure the assumption of temperature independence of the enthalpy.

Thermodynamic Network Calculations

The thermodynamic formation properties can be used to obtain the thermodynamic properties of a reaction, or the reaction data can be used to obtain the thermodynamic formation properties (Méndez 2008, Goldberg 2010). Such calculations are schematically represented in Figure 1.

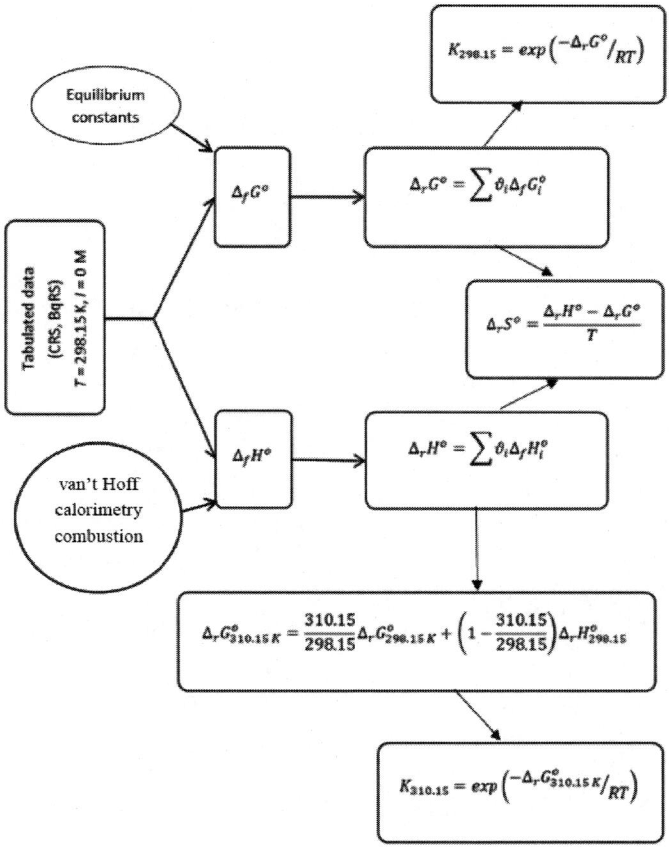

Figure 1. Thermodynamic network calculations from experimental and tabulated data, considering the standard state temperature of 298.15 K and the physiological temperature of 310.15 K.

Further Refinements on the Calculations

Once the thermodynamic properties are obtained in the CRS, it is possible to perform the first level of refinement of the data by making the calculations to change the reference state. The BRS, BqRS, and NPRS reference states are progressive refinements toward obtaining biochemically relevant data and the eqs. (10) to (14) allow those calculations to be made.

There are several methods to make these calculations. Prof. Robert A. Alberty led the first approaches to its realization. Based on the Legendre transforms (Alberty 1994a), Alberty defined new thermodynamic properties that he called *transformed properties* (Alberty 1992a), and that takes into account the description of the system under study under the conditions pH = 7, pMg = 3, and $I = 0$ M, defining the BqRS.

These calculations are difficult to perform because of the variety of simultaneous reactions that take place, including acid-base equilibria and complex formation equilibria with Mg^{2+} ions (Alberty 1992d, c). Its matrix nature requires the use of computers for calculations (Alberty 1991, 1992b).

Thermodynamic Properties and Thermodynamic Transformed Properties

Beyond the details and difficulties of the calculations, the work under the different standard states only requires thermodynamic tables in those reference states. In this way, all the thermodynamic calculations mentioned above can be performed in different reference states using thermodynamic properties (CRS) or transformed thermodynamic properties (BqRS). In the latter case, we will be in more relevant conditions from the biochemical point of view. As of its 97th edition, the *CRC Handbook of Chemistry and Physics* incorporates a table of transformed thermodynamic properties of biochemical compounds (Haynes 2016).

For BqRS, the species involved are no longer defined as charged species, but rather a set of related species (pseudoisomers) containing the species of interest. For example, phosphate at pH = 7 and pMg = 3, is represented as P_i and includes all species with different degrees of protonation and different degrees of complexation with Mg^{2+} ions, representative of these experimental conditions. In this way, the distinction between chemical and biochemical equations arises, establishing nomenclatures and rules defined by the IUBMB.

Adjustments to Physiological Conditions

Being able to count on values of thermodynamic properties in the real conditions in which a biochemical system operates is the ultimate goal of biochemical thermodynamics. From what we have seen so far, the necessary equations are available to carry out these calculations, although the property data in the different standard states, in particular, in the biochemical standard state, are not always available.

Two aspects are key here: the calculation of the Gibbs energy outside of equilibrium conditions, and the adequate description of the medium in which the biochemical process under study takes place. Out of equilibrium conditions, the reaction quotient (Q) relates to the Gibbs energy of reaction:

$$\Delta_r G = \Delta_r G^o + RT\ lnQ \tag{23}$$

and the sign of $\Delta_r G$ determines the spontaneous direction of the reaction. Note that although the equation is written for the CRS, it is valid for any reference state. The value of Q is determined experimentally, provided that analytical methodologies are available that allow the detection and quantification of the metabolites involved. These measurements are made under certain conditions that try to simulate the physiological environment as realistically as possible. An example of the calculation of Q is given by the table of absolute concentrations for more than 100 metabolites in *Escherichia coli* (Bennett et al. 2009).

Normal physiological conditions for most organisms involve temperatures between 37 and 38°C (Bergman, Kashiwaya, and Veech 2010), intracellular pH between 6 and 7.5 (Goldberg, Tewari, and Bhat 2004), free magnesium ion concentration between 0.2 and 2.5 mM (Günther 2006), and ionic strength between 0.15 M and 0.4 M. For typical mammalian cells, the following values are suggested: T = 310.15 K, I = 0.18 M, pH = 7, $[Mg^{2+}]$ = 0.8 mM, $[K^+]$ = 140 mM, $[Na^+]$ = 10 mM, $[Ca^{2+}]$ = 0.1 μM (Li et al. 2010). Finally, as a general suggestion to represent simplifyied physiological conditions, T = 310.15 K, pH = 7, pMg = 3, and I = 0.25 M are assumed (Goldberg, Tewari, and Bhat 2004).

Thermodynamic Tables for the Analysis of ATP Hydrolysis and the Glycolytic Pathway

In the CRS, the chemical equation that describes the process of ATP hydrolysis is:

$$Mg_2ATP\ (ac) + H_2O\ (ac) \rightleftharpoons MgADP^-\ (ac) + MgHPO_4\ (ac) + H^+\ (ac)$$

and the thermodynamic properties of the reaction can be calculated from the formation properties of each of the reactants and products (Table 4).

In the BqRS, the biochemical equation for this process is:

$$ATP + H_2O = ADP + P_i$$

where each symbol includes all pseudoisomers of the said species under fixed pH and pMg conditions.

The glycolytic pathway consists of ten enzyme-catalyzed reactions, which are represented by the following biochemical equations (see Table 5 for nomenclature):

1. Glu + ATP = G6P + ADP	6. GAP + NAD + Pi = BPG + NADH
2. G6P = F6P	7. BPG + ADP = 3PG + ATP
3. FDP = G3P + DHAP	8. 3PG = 2PG
4. F6P + ATP = FDP + ADP	9. 2PG = PEP
5. DHAP = G3P	10. PEP + ADP = Pyr + ATP

Table 4. Data availability (CRS) for the analysis of ATP hydrolysis and the glycolytic pathway (1 bar, 298.15 K, $I = 0$ M)[a]

Compound	Chemical acronym	$\Delta_f H_m^o$ (kJ mol^{-1})	$\Delta_f G_m^o$ (kJ mol^{-1})	S_m^o (J mol^{-1} K^{-1})
Adenosine 5'–diphosphate	ADP^{3-}	–2626.54	–1906.13	207.4
	$HADP^{2-}$	–2620.94	–1947.10	363.6
	H_2ADP^-	–2638.54	–1971.98	388.0
	$MgADP^-$	–3074.54	–2388.06	223.1
	MgHADP	–3074.54	–2416.67	316.3
Adenosine 5'–monophosphate	AMP^{2-}	–1635.37	–1040.45	214.5
	HAMP–	–1629.97	–1078.86	361.4
	H_2AMP	–1648.07	–1101.63	377.1
	MgAMP	–2091.07	–1511.77	168.8
Adenosine 5'–triphosphate	ATP^{4-}	–3619.21	–2768.10	182.8
	$HATP^{3-}$	–3612.91	–2811.48	349.4
	H_2ATP^{2-}	–3627.91	–2838.18	388.7
	$MgATP^{2-}$	–4063.31	–3258.77	340.9
	$MgHATP^-$	–4063.01	–3287.59	338.5
	Mg_2ATP	–4519.51	–3729.52	191.6
1,3–Biphosphoglycerate	BPG^{4-}		–2356.14	
	$HBPG^{3-}$		–2401.58	
	H_2BPG^{2-}		–2436.29	
	$MgBPG^{2-}$		–2836.07	
Dihydroxyacetone phosphate	$DHAP^{2-}$		–1296.26	
	$HDHAP^-$		1329.80	
	MgDHAP		–1770.0	
Fructose–1,6–diphosphate	$F16P^{4-}$		–2601.40	–225.9
	$HF16P^{3-}$		–2639.36	–92.6
	H_2F16P^2		–2673.89	–101.4
	$MgF16P^2$		–3080.00	
Fructose–6–phosphate	$F6P^{2-}$	–2264.77	–1760.80	156.7
	$HF6P^-$		–1796.60	282.5
	MgF6P		–2230.20	

Compound	**Chemical acronym**	$\Delta_f H^o_m$ (kJ mol^{-1})	$\Delta_f G^o_m$ (kJ mol^{-1})	S^o_m (J mol^{-1} K^{-1})
Glucose	GLC(cr)	–1268.10	–908.90	228.0
	GLC(ac)	–1262.19	–915.90	272.5
Glucose–6–phosphate	$G6P^{2-}$	–2276.44	–1763.94	128.1
	$HG6P^{-}$	–2274.64	–1800.59	258.1
	MgG6P	–2732.04	–2234.08	
Glyceraldehyde–3–phosphate	GAP^{2-}		–1288.60	
Glycerol–3–phosphate	$G3P^{2-}$		–1288.60	35.5
	$HG3P^{-}$		–1321.14	169.2
	MgG3P		–1756.44	
NAD	$HNAD^{-}_{red}$	–3274.51	–2207.65	149.6
	$NADH_{2\ red}$	–3266.56	–2216.43	163.0
	$NADH_2^{+}{}_{ox}$	–3237.02	–2238.90	173.5
	$HNAD_{ox}$	–3244.97	–2230.11	669.4
	$NAD^{-}{}_{ox}$	–3247.06	–2216.73	617.6
Phosphate	PO_4^{3-}	–1281.43	–1025.32	–220.3
	HPO_4^{2-}	–1299.00	–1096.10	–33.1
	$H_2PO_4^{-}$	–1303.14	–1137.20	92.5
	H_3PO_4	–1295.28	–1149.30	
	$MgHPO_4$	–1753.80	–1566.17	
2–phosphoglycerate	$PG2^{3-}$		–1496.38	
	$HPG2^{2-}$		–1539.99	179.5
	H_2PG^{2-}		–1562.67	259.8
	$MgPG^{2-}$		–1973.80	70.0
3–phosphoglycerate	$PG3^{3-}$		–1502.54	
	$HPG3^{2-}$		–1545.52	
	H_2PG^{3-}		–1564.27	
	$MgPG^{3-}$		–1977.28	
Phosphoenolpyruvate	PEP^{3-}	–1577.40	–1271.40	–25.7
	$HPEP^{2-}$	–1561.40	–1311.30	161.8
	H_2PEP^{-}	–1534.40	–1333.10	325.6
	$MgPEP^{-}$	–2022.40	–1746.50	–22.9
Pyruvate	Pyr^{-}	–574.88	–475.30	252.7
	HPyr	–587.02	–489.53	259.8
Water	H_2O(g)	–241.80	–228.60	–188.8

[a] Data were taken from (Alberty 2005, Alberty 2006a, Boerio-Goates et al. 2001, Canovas and Shock 2016, Goldberg and Tewari 2003, Ould-Moulaye, Dussap, and Gros 2002, Rajendran and Muthukumarasamy 2018).

Table 5. Data availability (BqRS) for the analysis of ATP hydrolysis and the glycolytic pathway (1 bar, 298.15 K, pH = 7, pMg = 3, I = 0 M and I = 0.25 M)[a]

Compound	**Biochemical acronym**	$\Delta_f H_m'^o$ **(kJ mol⁻¹)**		$\Delta_f G_m'^o$**(kJ mol⁻¹)**	
		I **= 0 M**	I **= 0.25 M**	I **= 0 M**	I **= 0.25 M**
Adenosine 5'–diphosphate	ADP	–2623.20	–2627.24	–1428.93	–1425.96
Adenosine 5'–monophosphate	AMP			–562.04	–554.83
Adenosine 5'–triphosphate	ATP	–3614.23	–3616.92	–2292.61	–2297.34
1,3–biphosphoglycerate	BPG			–2202.06	–2207.03
Dihydroxyacetone phosphate	DHAP			–1096.60	–1095.70
Fructose–1,6–diphosphate	FDP			–2202.84	–2206.78
Fructose–6–phosphate	F6P			–1321.71	–1315.74
Glucose	Glu			–436.42	–426.71
Glucose–6–phosphate	G6P			–1325.00	–1318.92
Glyceraldehyde–3–phosphate	GAP			–1088.94	–1088.04
Glycerol–3–phosphate	G3P			–1080.22	–1077.13
NAD oxidized	NADox		–10.26	1038.86	1059.10
NAD reduced	NADred		–41.38	1101.47	1120.09
Phosphate	Pi	–1301.24	–1299.36	–1058.56	–1047.17
2–Phosphoglycerate	2PG			–1340.72	–1341.79
3–Phosphoglycerate	3PG			–1346.38	–1347.73
Phosphoenolpyruvate	PEP			–1185.46	–1189.73
Pyruvate	Pyr			–352.40	–350.78
Water	H_2O	–285.83	–286.65	–157.28	–155.66

[a] Data were taken from (Alberty 2005, Alberty 2006a).

Alternative Methods

The calculations carried out by the Alberty method have allowed obtaining values of thermodynamic properties of a significant number of biochemical reactions, under relevant conditions. However, if the total number of known enzymatic reactions is considered, the data tabulated so far do not seem to be significant. For this reason, alternative calculation methods have been proposed to obtain the thermodynamic properties theoretically.

Group-Contribution Methods

Unknown formation Gibbs properties can be estimated quite accurately using group-contribution methods. These methods are based on the assumption that every compound can be decomposed in different chemical groups, each one contributing with a specific Gibbs energy of formation. Being the Gibbs energy an additive property, the total value corresponding to each compound will be the sum of the partial Gibbs energies:

$$\Delta_f G^o_{compound} = \sum_{group} N_{group} \Delta_f G^o \quad (24)$$

This approach has been used in chemical thermodynamics since its proposal (Benson and Buss 1958). Tabulated values for each chemical group represents the mean value obtained from different compounds, reflecting also the incidence of the chemical environment for such group (Benson et al. 1969).

Mavrovouniotis was the first to extend the use of group-contribution methods to biochemical compounds (Mavrovouniotis 1990, Mavrovouniotis et al. 1988) and biotransformations in general (Mavrovouniotis 1991), and Alberty proved that transformed thermodynamic properties could also be included (Alberty 1998). The original contribution estimates the Gibbs energy by decomposing compounds into their functional groups according to their biochemical functions. The original tables were later improved by the addition of new groups and interaction factors (Jankowski et al. 2008). New accessible thermodynamic data considered the influence of pH and ionic strength by including the different degrees of protonation of the compounds (Noor et al. 2012). Subsequently, the tables were completed with the introduction of entropy and enthalpy data (Du et al. 2018).

Component-Contribution Methods

One way to increase the database and improve predictions is to consider the part of the molecule that has changed in a biochemical reaction. Instead of group contributions, component contributions are considered (Noor et al. 2013).

Chemical Similarity Method

A new method of predicting Gibbs energy values starts from the assumption that similar chemical reactions should have similar Gibbs energies (Rother et al. 2010). In this way, the problem is limited to finding those chemical descriptors that allow comparing reactions to assign them the same Gibbs energy values. This idea was incorporated into a machine learning method (Alazmi et al. 2018) with improved prediction accuracy.

Quantum Calculations

Together with the group-contribution methods, the methods based on quantum mechanics constitute one of the main tools for the calculation of the Gibbs energies. The main goal of these methods is to achieve a chemical accuracy of 1 kcal mol^{-1}, i.e., 4.184 kJ mol^{-1} (van Speybroeck, Gani, and Meier 2010).

The prediction of Gibbs energies for biochemical reactions has to face several problems not found in gas-phase systems. Biochemical reactions take place in an aqueous medium, and as we have already mentioned, the pH exerts a huge influence on the level of protonation of the species involved, and the charged species may require additional methods for accurate description.

Calculations based on the Density Functional Theory (DFT), using the functional B3LYP and the 6-31G* basis set yield Gibbs energy values for the different metabolites with different degrees of protonation, which are then considered together using Legendre transformation methods to study the effect of pH, and thus obtain the transformed thermodynamic properties. With this procedure, absolute errors varied in the range of 0.4 to 80 kJ mol^{-1} (Jinich et al. 2014).

Recently, DFT calculations were optimized using different basis sets, solvation models, and exchange-correlation functionals for the calculation of Gibbs energies (Joshi et al. 2021). The results were compared with the experimental values of about 300 biochemical reactions, reaching absolute errors in the range of 6.7 - 9.5 kJ mol^{-1}. Although the error is close to the ideal expected for this type of calculation, a bias is still observed when experimental and calculated data are compared.

Web-Based Calculators and Software

In addition to the lack of thermodynamic data for metabolites, there is an enormous amount of computational work that needs to be done to calculate the thermodynamic properties under relevant biochemical conditions. Given this, web platforms have been developed with friendly interfaces that facilitate these calculations.

eQuilibrator is a web platform that was developed to facilitate the calculations for a greater spectrum of users. It allows the calculation of the thermodynamic properties of compounds and reactions at arbitrary values of pH, ionic strength, and metabolite concentrations (Flamholz et al. 2012). It has an open-source code and access to the original data. This platform allows obtaining the thermodynamic data of interest without going through the work of its manual calculation, which provides greater accessibility for its use. The Gibbs energies of formation of the compounds in the database are those calculated by Alberty (Alberty 2005), or those obtained by the group contribution method (Noor et al. 2012, Noor et al. 2013). Some interesting examples in the literature (Blaß, Weyler, and Heinzle 2017, Halling 2020) demonstrate the importance of this platform in the analysis of theoretical reactions.

The latest version eQuilibrator.3.0, (equilibrator.weizmann.ac.il), incorporates an Application Programming Interface that can be integrated with other commonly used tools such as the Kyoto Encyclopedia of Genes and Genomes (KEGG) database, IUPAC International Chemical Identifier (InChI), and Simplified Molecular-Input Line-Entry System (SMILES) (Beber et al. 2021). The new platform allows new compounds to be entered, takes into account corrections due to the presence of Mg^{2+} ions, and considerably increases its database.

Group contribution methods do not cover all metabolites, nor do they consider stereoisomers, predicting null values for isomerase-catalyzed reactions. dGPredictor uses training datasets and machine learning algorithms to generate chemical groups, interactions, and distinguish stereoisomers in Gibbs energy calculations (Wang, Upadhyay, and Maranas 2021).

On the Reunification of Chemical and Biochemical Thermodynamics

In 2010, Iotti's group proposed that chemical and biochemical thermodynamics should be able to be brought together, based on the fact that, by both experimental measurements and theoretical proofs, the Gibbs energies under non-equilibrium conditions were the same in both systems (Iotti, Sabatini, and Vacca 2010). In that same work, they also show that the equality occurs at the level of the Gibbs energies out of equilibrium ($\Delta_r G = \Delta_r G'$), but that the inequality persists for other thermodynamic properties: $\Delta_r G^o \neq \Delta_r G'^o$, $\Delta_r H \neq \Delta_r H'$, $\Delta_r S \neq \Delta_r S'$.

According to the proposal, the values of the transformed thermodynamic properties can be obtained directly by balancing the reactions written as chemical (not biochemical) equations and calculating the stoichiometric coefficients for the pseudoisomers (Sabatini, Vacca, and Iotti 2012). In this way, the need to apply Legendre transforms, which require more complex computational methods, is avoided, and the use of an electronic spreadsheet is sufficient (Iotti, Raff, and Sabatini 2017). This procedure is now known as Balanced Biochemical Reaction (BBR) method.

A working group was created in 2017 to study the possible reunification of chemical and biochemical thermodynamics. The objective was "to show that the two worlds of chemical and biochemical thermodynamics, which so far have been treated separately, can be reunified within the same thermodynamic framework," and the project ended in 2021 with the publication of a Technical Report (Sabatini et al. 2021).

A recent study discusses the validity of the BBR proposal and that of Alberty on Legendre transforms (Cannon and Raff 2021). Regarding the BBR method, they conclude that it does not give the same results as those obtained by Alberty, not fulfilling the $\Delta_r G = \Delta_r G'$ condition. They indicate that the differences in the calculations are small (in the order of 2.0 kJ mol^{-1}), and arise from the consideration or not of the pseudoisomers in the equilibrium. The authors also indicate that both methods are correct in their formulation and that their application would depend on the experimental or *in vivo* conditions applied. About the Alberty method, the authors conclude that it is not necessary to obtain a new state function based on a Legendre transformation to fix the pH value and provide alternative calculation methods.

Conclusion

Thermodynamics applied to biochemical systems is a relatively new discipline and is in continuous transformation. The incidence of physiological conditions in the values of thermodynamic properties is of such relevance that it has led to the creation of biochemical thermodynamics as a new branch of Physical Chemistry.

The calculations of the thermodynamic properties for the conditions relevant to biochemistry are challenging due to the complexity of the calculations themselves and the scarcity of experimental data. Group-contribution methods and quantum chemistry calculations help to fill the gap, but it gives rise to data tables that are incompatible with the existing ones.

The recent proposal to unify biochemical and chemical thermodynamics has been taken up as a joint working project by IUPAC and IUBMB. However, the data and inferences still fall short of clearly establishing whether this is possible.

Nearly 100 years after the first experimental measurement of an enzyme-catalyzed reaction, biochemical thermodynamics is under scrutiny and its history is still being written.

Acknowledgments

To Dr. Santiago Botasini for fruitful discussions on the draft of the manuscript.

References

Alberty, R. A., and R. N. Goldberg. 1994. "Calorimetric determination of the standard transformed enthalpy of a biochemical reaction at specified pH and pMg (Biophysical Chemistry 47 (1993) 213-223)." *Biophysical Chemistry* no. 49 (2):184. doi: 10.1016/0301-4622(94)85003-8.

Acree, William E., and James S. Chickos. 2021. "JPCRD: 50 Years of Providing the Scientific Community with Critically Evaluated Thermodynamic Data, Predictive Methods, and Large Thermodynamic Data Compilations." *Journal of Physical and Chemical Reference Data* no. 50 (3):033101. doi: 10.1063/5.0057525.

Alazmi, Meshari, Hiroyuki Kuwahara, Othman Soufan, Lizhong Ding, and Xin Gao. 2018. "Systematic selection of chemical fingerprint features improves the Gibbs energy prediction of biochemical reactions." *Bioinformatics* no. 35 (15):2634-2643. doi: 10.1093/bioinformatics/bty1035.

Alberty, R. A. 1991. "Chemical equations are actually matrix equations." *Journal of Chemical Education* no. 68 (12):984. doi: 10.1021/ed068p984.

Alberty, R. A. 1992a. "Calculation of transformed thermodynamic properties of biochemical reactants at specified pH and pMg." *Biophysical Chemistry* no. 43 (3):239-254. doi: 10.1016/0301-4622(92)85024-X.

Alberty, R. A. 1992b. "Conversion of chemical equations to biochemical equations." *Journal of Chemical Education* no. 69 (6):493. doi: 10.1021/ed069p493.

Alberty, R. A. 1992c. "Degrees of freedom in biochemical reaction systems at specified pH and pMg." *Journal of Physical Chemistry* no. 96 (24):9614-9621. doi: 10.1021/j100203a012.

Alberty, R. A. 1992d. "Equilibrium calculations on systems of biochemical reactions at specified pH and pMg." *Biophysical Chemistry* no. 42 (2):117-131. doi: 10.1016/0301-4622(92)85002-L.

Alberty, R. A. 1994a. "Legendre transforms in chemical thermodynamics." *Chemical Reviews* no. 94 (6):X-1482.

Alberty, R. A. 1994b. "Recommendations for nomenclature and tables in biochemical thermodynamics (IUPAC Recommendations 1994)." *Pure and Applied Chemistry* no. 66 (8):1641-1666. doi: 10.1351/pac199466081641.

Alberty, R. A. 1995. "Standard transformed formation properties of carbon dioxide in aqueous solutions at specified pH." *Journal of Physical Chemistry* no. 99 (27):11028-11034. doi: 10.1021/j100027a051.

Alberty, R. A. 1998. "Calculation of standard transformed entropies of formation of biochemical reactants and group contributions at specified pH." *Journal of Physical Chemistry A* no. 102 (44):8460-8466. doi: 10.1021/jp981003p.

Alberty, R. A. 2004. "A short history of the thermodynamics of enzyme-catalyzed reactions." *Journal of Biological Chemistry* no. 279 (27):27831-27836. doi: 10.1074/jbc.X400003200.

Alberty, R. A. 2006a. Biochemical Thermodynamics: Applications of Mathematica. In *Methods of Biochemical Analysis*.

Alberty, R. A. 2006b. "Standard molar entropies, standard entropies of formation, and standard transformed entropies of formation in the thermodynamics of enzyme-catalyzed reactions." *Journal of Chemical Thermodynamics* no. 38 (4):396-404. doi: 10.1016/j.jct.2005.06.006.

Alberty, R. A. 2007. "Changes in the standard transformed thermodynamic properties of enzyme-catalyzed reactions with ionic strength." *Journal of Physical Chemistry B* no. 111 (14):3847-3852. doi: 10.1021/jp068883p.

Alberty, R. A., A. Cornish-Bowden, R. N. Goldberg, G. G. Hammes, K. Tipton, and H. V. Westerhoff. 2011. "Recommendations for terminology and databases for biochemical thermodynamics." *Biophysical Chemistry* no. 155 (2-3):89-103. doi: 10.1016/j.bpc.2011.03.007.

Alberty, R. A., and R. N. Goldberg. 1993. "Calorimetric determination of the standard transformed enthalpy of a biochemical reaction at specified pH and pMg." *Biophysical Chemistry* no. 47 (3):213-223. doi: 10.1016/0301-4622(93)80046-L.

Alberty, R.A. 2005. *Thermodynamics of Biochemical Reactions*: Wiley.

Alberty, Robert A. 2010. "Brief Scientific Autobiography of Robert A. Alberty." *The Journal of Physical Chemistry B* no. 114 (49):16047-16050. doi: 10.1021/jp103554e.

Beber, M. E., M. G. Gollub, D. Mozaffari, K. M. Shebek, A. I. Flamholz, R. Milo, and E. Noor. 2021. "eQuilibrator 3.0: a database solution for thermodynamic constant estimation." *Nucleic Acids Res.* doi: 10.1093/nar/gkab1106.

Bennett, B. D., E. H. Kimball, M. Gao, R. Osterhout, S. J. Van Dien, and J. D. Rabinowitz. 2009. "Absolute metabolite concentrations and implied enzyme active site occupancy in Escherichia coli." *Nat Chem Biol* no. 5 (8):593-9. doi: 10.1038/nchembio.186.

Benson, Sidney W., and Jerry H. Buss. 1958. "Additivity Rules for the Estimation of Molecular Properties. Thermodynamic Properties." *The Journal of Chemical Physics* no. 29 (3):546-572. doi: 10.1063/1.1744539.

Benson, Sidney W., F. R. Cruickshank, D. M. Golden, Gilbert R. Haugen, H. E. O'Neal, A. S. Rodgers, Robert Shaw, and R. Walsh. 1969. "Additivity rules for the estimation of thermochemical properties." *Chemical Reviews* no. 69 (3):279-324. doi: 10.1021/cr60259a002.

Bergman, Christian, Yoshihiro Kashiwaya, and Richard L. Veech. 2010. "The Effect of pH and Free Mg2+ on ATP Linked Enzymes and the Calculation of Gibbs Free Energy of ATP Hydrolysis." *The Journal of Physical Chemistry B* no. 114 (49):16137-16146. doi: 10.1021/jp105723r.

Blaß, L. K., C. Weyler, and E. Heinzle. 2017. "Network design and analysis for multi-enzyme biocatalysis." *BMC Bioinformatics* no. 18 (1). doi: 10.1186/s12859-017-1773-y.

Boerio-Goates, Juliana, Michael R. Francis, Robert N. Goldberg, Manuel A. V. Ribeiro da Silva, Maria D. M. C. Ribeiro da Silva, and Yadu B. Tewari. 2001. "Thermochemistry of adenosine." *The Journal of Chemical Thermodynamics* no. 33 (8):929-947. doi: 10.1006/jcht.2001.0820.

Burton, K., and H. A. Krebs. 1953. "The free-energy changes associated with the individual steps of the tricarboxylic acid cycle, glycolysis and alcoholic fermentation and with the hydrolysis of the pyrophosphate groups of adenosinetriphosphate." *Biochemical Journal* no. 54 (1):94-107. doi: 10.1042/bj0540094.

Cannon, W. R., and L. M. Raff. 2021. "The formulation of chemical potentials and free energy changes in biochemical reactions." *Phys Chem Chem Phys* no. 23 (27):14783-14795. doi: 10.1039/d1cp02045e.

Canovas, Peter A., and Everett L. Shock. 2016. "Geobiochemistry of metabolism: Standard state thermodynamic properties of the citric acid cycle." *Geochimica et Cosmochimica Acta* no. 195:293-322. doi: 10.1016/j.gca.2016.08.028.

Cornish-Bowden, Athel, Robert N. Goldberg, and Gordon G. Hammes. 2010. "Professor Robert A. Alberty—A Legacy of Excellence." *The Journal of Physical Chemistry B* no. 114 (49):16045-16046. doi: 10.1021/jp105228e.

Du, B., Z. Zhang, S. Grubner, J. T. Yurkovich, B. O. Palsson, and D. C. Zielinski. 2018. "Temperature-Dependent Estimation of Gibbs Energies Using an Updated Group-Contribution Method." *Biophys J* no. 114 (11):2691-2702. doi: 10.1016/j.bpj.2018.04.030.

Emel'yanenko, Vladimir N., Dzmitry H. Zaitsau, and Sergey P. Verevkin. 2017. "Thermochemical Properties of Xanthine and Hypoxanthine Revisited." *Journal of Chemical & Engineering Data* no. 62 (9):2606-2609. doi: 10.1021/acs.jced.7b00085.

Engel, T., and P. Reid. 2020. *Physical Chemistry: Thermodynamics, Statistical Thermodynamics, and Kinetics*: Pearson Publishing.

Flamholz, A., E. Noor, A. Bar-Even, and R. Milo. 2012. "eQuilibrator--the biochemical thermodynamics calculator." *Nucleic Acids Res* no. 40 (Database issue):D770-5. doi: 10.1093/nar/gkr874.

Fuchs, H., and R. Gessner. 2001. "The result of equilibrium-constant calculations strongly depends on the evaluation method used and on the type of experimental errors." *The Biochemical journal* no. 359 (Pt 2):411-418. doi: 10.1042/0264-6021:3590411.

Goldberg, R. N. 1999. "Thermodynamics of enzyme-catalyzed reactions: Part 6 - 1999 update." *Journal of Physical and Chemical Reference Data* no. 28 (4):931-965. doi: 10.1063/1.556041.

Goldberg, R. N., and Y. B. Tewari. 1989. "A calorimetric and equilibrium investigation of the hydrolysis of lactose." *Journal of Biological Chemistry* no. 264 (17):9897-9900. doi: 10.1016/s0021-9258(18)81743-4.

Goldberg, R. N., and Y. B. Tewari. 1994a. "Thermodynamics of Enzyme-Catalyzed Reactions. Part 3. Hydrolases." *Journal of Physical and Chemical Reference Data* no. 23 (6):1035-1103. doi: 10.1063/1.555957.

Goldberg, R. N., and Y. B. Tewari. 1994b. "Thermodynamics of Enzyme-Catalyzed Reactions: Part 2. Transferases." *Journal of Physical and Chemical Reference Data* no. 23 (4):547-617. doi: 10.1063/1.555948.

Goldberg, R. N., and Y. B. Tewari. 1995a. "Thermodynamics of Enzyme-Catalyzed Reactions: Part 4. Lyases." *Journal of Physical and Chemical Reference Data* no. 24 (5):1669-1698. doi: 10.1063/1.555969.

Goldberg, R. N., and Y. B. Tewari. 1995b. "Thermodynamics of Enzyme-Catalyzed Reactions: Part 5. Isomerases and Ligases." *Journal of Physical and Chemical Reference Data* no. 24 (6):1765-1801. doi: 10.1063/1.555970.

Goldberg, R. N., and Y. B. Tewari. 2003. "Thermodynamics of the hydrolysis reactions of adenosine 3′,5′-(cyclic)phosphate(aq) and phosphoenolpyruvate(aq); the standard molar formation properties of 3′,5′-(cyclic)phosphate(aq) and phosphoenolpyruvate(aq)." *Journal of Chemical Thermodynamics* no. 35 (11):1809-1830. doi: 10.1016/j.jct.2003.08.002.

Goldberg, R. N., Y. B. Tewari, and J. C. Ahluwalia. 1989. "Thermodynamics of the hydrolysis of sucrose." *Journal of Biological Chemistry* no. 264 (17):9901-9904. doi: 10.1016/s0021-9258(18)81744-6.

Goldberg, R. N., Y. B. Tewari, D. Bell, K. Fazio, and E. Anderson. 1993. "Thermodynamics of Enzyme-Catalyzed Reactions: Part 1. Oxidoreductases." *Journal of Physical and Chemical Reference Data* no. 22 (2):515-582. doi: 10.1063/1.555939.

Goldberg, R. N., Y. B. Tewari, and T. N. Bhat. 2004. "Thermodynamics of enzyme-catalyzed reactions--a database for quantitative biochemistry." *Bioinformatics* no. 20 (16):2874-7. doi: 10.1093/bioinformatics/bth314.

Goldberg, R. N., Y. B. Tewari, and T. N. Bhat. 2007. "Thermodynamics of enzyme-catalyzed reactions: Part 7-2007 update." *Journal of Physical and Chemical Reference Data* no. 36 (4):1347-1397. doi: 10.1063/1.2789450.

Goldberg, Robert N. 2010. Thermodynamic Network Calculations Applied to Biochemical Substances and Reactions, 2010/09/14/.

Goldberg, Robert N. 2014. "Standards in biothermodynamics." *Perspectives in Science* no. 1 (1):7-14. doi: https://doi.org/10.1016/j.pisc.2014.02.001.

Günther, T. 2006. "Concentration, compartmentation and metabolic function of intracellular free Mg2+." *Magnes Res* no. 19 (4):225-36.

Halling, P. J. 2020. "Thermodynamic Favorability of End Products of Anaerobic Glucose Metabolism." *ACS Omega* no. 5 (26):15843-15849. doi: 10.1021/acsomega.0c00790.

Haynes, W. M. 2016. *CRC Handbook of Chemistry and Physics*. 97th. ed: CRC Press.

Hinz, H. J., and H. Hans-Jurgen. 1986. *Thermodynamic Data for Biochemistry and Biotechnology*: Springer-Verlag.

Iotti, S., L. Raff, and A. Sabatini. 2017. "Chemical and biochemical thermodynamics: Is it time for a reunification?" *Biophys Chem* no. 221:49-57. doi: 10.1016/j.bpc.2016.10.004.

Iotti, S., A. Sabatini, and A. Vacca. 2010. "Chemical and biochemical thermodynamics: from ATP hydrolysis to a general reassessment." *J Phys Chem B* no. 114 (5):1985-93. doi: 10.1021/jp903990j.

Jankowski, Matthew D., Christopher S. Henry, Linda J. Broadbelt, and Vassily Hatzimanikatis. 2008. "Group Contribution Method for Thermodynamic Analysis of Complex Metabolic Networks." *Biophysical Journal* no. 95 (3):1487-1499. doi: 10.1529/biophysj.107.124784.

Jin, C., X. Zhu, H. Wu, Y. Wang, and X. Hu. 2020. "Perturbation of phosphoglycerate kinase 1 (PGK1) only marginally affects glycolysis in cancer cells." *Journal of Biological Chemistry* no. 295 (19):6425-6446. doi: 10.1074/jbc.RA119.012312.

Jinich, A., D. Rappoport, I. Dunn, B. Sanchez-Lengeling, R. Olivares-Amaya, E. Noor, A. B. Even, and A. Aspuru-Guzik. 2014. "Quantum chemical approach to estimating the thermodynamics of metabolic reactions." *Scientific Reports* no. 4. doi: 10.1038/srep07022.

Joshi, R. P., A. McNaughton, D. G. Thomas, C. S. Henry, S. R. Canon, L. A. McCue, and N. Kumar. 2021. "Quantum Mechanical Methods Predict Accurate Thermodynamics of Biochemical Reactions." *ACS Omega* no. 6 (14):9948-9959. doi: 10.1021/acsomega.1c00997.

Krebs, H.A., H.L. Kornberg, S.N. Blagoveshchenskii, and K. Burton. 1957. *Energy Transformations in Living Matter: A Survey*: Springer-Verlag.

Li, X., R. K. Dash, R. K. Pradhan, F. Qi, M. Thompson, K. C. Vinnakota, F. Wu, F. Yang, and D. A. Beard. 2010. "A Database of Thermodynamic Quantities for the Reactions of Glycolysis and the Tricarboxylic Acid Cycle." *The Journal of Physical Chemistry B* no. 114 (49):16068-16082. doi: 10.1021/jp911381p.

Li, X., F. Wu, F. Qi, and D. A. Beard. 2011. "A database of thermodynamic properties of the reactions of glycolysis, the tricarboxylic acid cycle, and the pentose phosphate pathway." *Database* no. 2011. doi: 10.1093/database/bar005.

Lipmann, Fritz. 1941. "Metabolic Generation and Utilization of Phosphate Bond Energy." *Advances in Enzymology and Related Areas of Molecular Biology*:99-162. doi: https://doi.org/10.1002/9780470122464.ch4.

Mavrovouniotis, M. L. 1990. "Group contributions for estimating standard gibbs energies of formation of biochemical compounds in aqueous solution." *Biotechnology and Bioengineering* no. 36 (10):1070-1082. doi: https://doi.org/10.1002/bit.260361013.

Mavrovouniotis, M. L. 1991. "Estimation of standard Gibbs energy changes of biotransformations." *Journal of Biological Chemistry* no. 266 (22):14440-14445. doi: 10.1016/S0021-9258(18)98705-3.

Mavrovouniotis, Michael L., Patrick Bayol, Tu-Kien Michael Lam, George Stephanopoulos, and Gregory Stephanopoulos. 1988. "A group contribution method for the estimation of equilibrium constants for biochemical reactions." *Biotechnology Techniques* no. 2 (1):23-28. doi: 10.1007/BF01874203.

Mendez, E. 2020. "Choosing the adequate thermodynamic reference state: The evolution of pyrophosphate bioenergetics as an example." *Biochem Mol Biol Educ* no. 48 (3):247-252. doi: 10.1002/bmb.21333.

Méndez, E., and M. F. Cerdá. 2016. "Discovering Reliable Sources of Biochemical Thermodynamic Data to Aid Students Understanding." *Journal of Chemical Education* no. 93 (3):555-559. doi: 10.1021/acs.jchemed.5b00412.

Méndez, Eduardo. 2008. "Biochemical thermodynamics under near-physiological conditions." *Biochemistry and Molecular Biology Education* no. 36 (2):116-119. doi: https://doi.org/10.1002/bmb.20157.

Miller, Stanley L., and David Smith-Magowan. 1990. "The Thermodynamics of the Krebs Cycle and Related Compounds." *Journal of Physical and Chemical Reference Data* no. 19 (4):1049-1073. doi: 10.1063/1.555878.

Noor, E., A. Bar-Even, A. Flamholz, L. Lubling, D. Davidi, and R. Milo. 2012. "An integrated open framework for thermodynamics of reactions that combines accuracy and coverage." *Bioinformatics* no. 28 (15):2037-2044. doi: 10.1093/bioinformatics/bts317.

Noor, E., H. S. Haraldsdóttir, R. Milo, and R. M. T. Fleming. 2013. "Consistent Estimation of Gibbs Energy Using Component Contributions." *PLoS Computational Biology* no. 9 (7). doi: 10.1371/journal.pcbi.1003098.

Ould-Moulaye, C. B., C. G. Dussap, and J. B. Gros. 2002. "A consistent set of formation properties of nucleic acid compounds: Nucleosides, nucleotides and nucleotide-phosphates in aqueous solution." *Thermochimica Acta* no. 387 (1):1-15. doi: https://doi.org/10.1016/S0040-6031(01)00814-0.

Quastel, Juda Hirsch, and Barnet Woolf. 1926. "The Equilibrium between l-Aspartic Acid, Fumaric Acid and Ammonia in Presence of Resting Bacteria." *Biochemical Journal* no. 20 (3):545-555. doi: 10.1042/bj0200545.

Rajendran, Tamil Elakkiya, and Thenmozhi Muthukumarasamy. 2018. "Thermodynamic calculations of biochemical reaction systems at specified pH, pMg, and change in binding of hydrogen and magnesium ions." *Asia-Pacific Journal of Chemical Engineering* no. 13 (4):e2205. doi: 10.1002/apj.2205.

Rother, Kristian, Sabrina Hoffmann, Sascha Bulik, Andreas Hoppe, Johann Gasteiger, and Herrmann-Georg Holzhütter. 2010. "IGERS: Inferring Gibbs Energy Changes of

Biochemical Reactions from Reaction Similarities." *Biophysical Journal* no. 98 (11):2478-2486. doi: 10.1016/j.bpj.2010.02.052.

Sabatini, Antonio, Marco Borsari, Gerard P. Moss, and Stefano Iotti. 2021. "Chemical and biochemical thermodynamics reunification (IUPAC Technical Report)." *Pure and Applied Chemistry* no. 93 (2):243-252. doi: 10.1515/pac-2019-0908.

Sabatini, Antonio, Alberto Vacca, and Stefano Iotti. 2012. "Balanced Biochemical Reactions: A New Approach to Unify Chemical and Biochemical Thermodynamics." *PLOS ONE* no. 7 (1):e29529. doi: 10.1371/journal.pone.0029529.

Tewari, Y. B., and R. N. Goldberg. 1989. "Thermodynamics of hydrolysis of disaccharides." *Journal of Biological Chemistry* no. 264 (7):3966-3971. doi: 10.1016/s0021-9258(19)84947-5.

Tewari, Y. B., D. K. Steckler, and R. N. Goldberg. 1988. "Thermodynamics of isomerization reactions involving sugar phosphates." *Journal of Biological Chemistry* no. 263 (8):3664-3669. doi: 10.1016/s0021-9258(18)68976-8.

Tewari, Y. B., D. K. Steckler, R. N. Goldberg, and W. L. Gitomer. 1988. "Thermodynamics of hydrolysis of sugar phosphates." *Journal of Biological Chemistry* no. 263 (8):3670-3675. doi: 10.1016/s0021-9258(18)68977-x.

Thauer, R. K., K. Jungermann, and K. Decker. 1977. "Energy conservation in chemotrophic anaerobic bacteria." *Bacteriol Rev* no. 41 (1):100-80. doi: 10.1128/br.41.1.100-180.1977.

van Speybroeck, Veronique, Rafiqul Gani, and Robert Johan Meier. 2010. "The calculation of thermodynamic properties of molecules." *Chemical Society Reviews* no. 39 (5):1764-1779. doi: 10.1039/B809850F.

Wang, L., V. Upadhyay, and C. D. Maranas. 2021. "dGPredictor: Automated fragmentation method for metabolic reaction free energy prediction and de novo pathway design." *PLoS Comput Biol* no. 17 (9):e1009448. doi: 10.1371/journal.pcbi.1009448.

Xie, J., C. Dai, and X. Hu. 2016. "Evidence that does not support pyruvate kinase M2 (PKM2) - Catalyzed reaction as a rate-limiting step in cancer cell glycolysis." *Journal of Biological Chemistry* no. 291 (17):8987-8999. doi: 10.1074/jbc.M115.704825.

Xie, J., H. Wu, C. Dai, Q. Pan, Z. Ding, D. Hu, B. Ji, Y. Luo, and X. Hu. 2014. "Beyond Warburg effect - Dual metabolic nature of cancer cells." *Scientific Reports* no. 4. doi: 10.1038/srep04927.

Xu, Jimmy, Julia Martien, Cole Gilbertson, Junyu Ma, Daniel Amador-Noguez, and Junyoung O. Park. 2020. "Metabolic flux analysis and fluxomics-driven determination of reaction free energy using multiple isotopes." *Current Opinion in Biotechnology* no. 64:151-160. doi: https://doi.org/10.1016/j.copbio.2020.02.018.

Zhu, X., C. Jin, Q. Pan, and X. Hu. 2021. "Determining the quantitative relationship between glycolysis and GAPDH in cancer cells exhibiting the Warburg effect." *Journal of Biological Chemistry* no. 296. doi: 10.1016/j.jbc.2021.100369.

Chapter 6

In-Vitro Activity of Ceftolozane/Tazobactam against *Pseudomonas Aeruginosa* and *Enterobacterales* Isolates from Taiwan

Yee-Huang Ku[1], Po-An Su[2], Yin-Ching Chuang[2] and Wen-Liang Yu[3,4,*]

[1]Division of Infectious Disease, Department of Internal Medicine, Chi Mei Medical Center, Liouying, Tainan, Taiwan
[2]Division of Infectious Diseases, Department of Medicine, Chi Mei Medical Center, Tainan City, Taiwan
[3]Department of Intensive Care Medicine, Chi Mei Medical Center, Tainan City, Taiwan
[4]Department of Medicine, School of Medicine, College of Medicine, Taipei Medical University, Taipei City, Taiwan

Abstract

Background. Ceftolozane/tazobactam (TOL/TAZ), a new antipseudomonal cephalosporin combined with a β-lactamase inhibitor, is indicated for the treatment of complicated intra-abdominal infections (requiring in combination with metronidazole), complicated urinary tract infections, and nosocomial or ventilator-associated pneumonia.

Methods. Activities of TOL/TAZ and ten comparators (cefepime, cefoperazone/sulbactam, ceftazidime, ciprofloxacin, colistin, imipenem, levofloxacin, meropenem, piperacillin/tazobactam [PIP/TAZ] and tigecycline) were evaluated against isolates collected in Taiwan. The susceptibility was tested by broth microdilution methods.

Results. A total of 599 isolates were collected during 2012-2015, including intra-abdomen (n = 223), respiratory tract (n = 153), urine (n = 149), skin and soft tissue (n = 55), blood (n = 11) and others (n = 8).

* Corresponding Author's Email: yu2231@gmail.com.

In: Advances in Biology. Volume 1
Editor: Charles D. Grant
ISBN: 979-8-88697-201-6

TOL/TAZ demonstrated a good activity against *Enterobacterales* isolates (*n* = 446; 84.8% susceptibility; susceptible breakpoint, ≤ 2 μg/mL). TOL/TAZ was highly active versus *Citrobacter koseri* (*n* = 15), *Klebsiella oxytoca* (*n* = 30), *Morganella morganii* (*n* = 15), *Proteus mirabilis* (*n* = 30), and *Escherichia coli* (*n* = 125), with MIC_{90} values of ≤ 1 μg/mL. TOL/TAZ had the best overall activity versus *P. aeruginosa* (94% susceptibility; MIC_{90} = 4 μg/mL; susceptible breakpoint, ≤4 μg/mL), in comparison to imipenem (59% susceptibility, MIC_{90} >8 μg/mL), meropenem (71% susceptibility, MIC_{90} >8 μg/mL), and colistin (100% intermediate, MIC_{90} = 2 μg/mL). The susceptibility rates of TOL/TAZ for *K. pneumoniae* isolates susceptible and non-susceptible to ceftazidime were 75.2% and 31.1%, respectively.

Conclusions. TOL/TAZ demonstrated potent *in-vitro* activity against *P. aeruginosa* and *Enterobacterales* isolates in Taiwan. TOL/TAZ is a good option for treatment of infections caused by these organisms.

Keywords: ceftolozane, ceftolozane/tazobactam, *Enterobacterales*, *Pseudomonas aeruginosa*

1. Introduction

Ceftolozane is a novel cephalosporin with a chemical structure similar to that of ceftazidime, carrying a modified side-chain at the 3-position of the cephem nucleus, which binds to the target of penicillin-binding proteins (PBPs) and confers potent antipseudomonal activity via the inhibition of bacterial cell wall biosynthesis [1, 2]. In contrast to other β-lactams, ceftolozane is a potent PBP3 inhibitor and has a higher affinity for PBP1b and PBP1c [2]. Ceftolozane also displays increased activity against *Enterobacterales*, including those that harbor classical β-lactamases (e.g., TEM-1 and SHV-1). Ceftolozane, similar to traditional oxyimino-cephalosporins such as ceftazidime and ceftriaxone, would be hydrolyzed by extended-spectrum β-lactamases (ESBLs) and carbapenemases [1]. The addition of the β-lactamase inhibitor tazobactam extends the activity of ceftolozane to include most ESBL and AmpC β-lactamase producers [1, 2]. Therefore, ceftolozane/tazobactam (TOL/TAZ) being a new antipseudomonal β-lactam/β-lactamase inhibitor combination has potent activity versus multidrug-resistant *P. aeruginosa* and *Enterobacterales*, including phenotypes resistant to ceftazidime, cefepime, piperacillin/ tazobactam, and even carbapenems of non-carbapenemase-producing mechanisms [1].

2. Epidemiology in Antimicrobial Suceptibility

Among 603 *P. aeruginosa* and 5950 *Enterobacterales* isolates from the urinary tract and intraabdominal infections in Europe, TOL/TAZ showed better activities against ESBL-phenotype *Enterobacterales* ($MIC_{50/90}$, 0.5/8 μg/mL) than those isolates with carbapenem resistance ($MIC_{50/90}$, >32/>32 μg/mL). TOL/TAZ inhibited 92% of *P. aeruginosa* isolates at an MIC of ≤ 4 μg/mL. These *P. aeruginosa* isolates exhibited high rates of non-susceptibility to ceftazidime (23%), cefepime (21%), piperacillin/tazobactam (27%), and meropenem (21%). TOL/TAZ was active against 61%-70% of the isolates with any of these four-drug resistant phenotypes and was the highest activity of any β-lactam tested [3].

With regard to isolates in the Unites States (U.S.), *P. aeruginosa* clinical isolates collected from 32 U.S. hospitals, the activity of TOL/TAZ (97% susceptible) was greater than those of cefepime (86%), ceftazidime (85%), meropenem (82%), and piperacillin-tazobactam (80%) [4]. Of the meropenem-nonsusceptible *P. aeruginosa* isolates, 88% were susceptible to TOL/TAZ [4]. Among 145 *Escherichia coli* and *Klebsiella pneumoniae* respiratory isolates harboring ESBL genes (mainly CTX-M genes) collected from patients hospitalized in the U. S., TOL/TAZ inhibited 83% of the ESBL producers and was more active than cefepime (only 15% susceptible) and piperacillin-tazobactam (74% susceptible). TOL/TAZ showed activity (85% susceptible) against AmpC producers but displayed limited activity against the carbpenemase-producing bla_{KPC}-carrying isolates. For *P. aeruginosa* groups, TOL/TAZ inhibited 78 to 85% of isolates nonsusceptible to antipseudomonal β-lactams and 87% of the multidrug-resistant isolates tested [5].

A similar activity of TOL/TAZ (91% susceptible) was found against 489 clinical *P. aeruginosa* isolates collected from patients at 14 medical centres in the Asia-Pacific region with healthcare-associated infections [6]. TOL/TAZ demonstrated a good *in-vitro* activity ($MIC_{50/90}$, 0.5/16 μg/mL) against ESBL-phenotype *Enterobacterales* without carbapenem-resistant phenotype [6]. While focusing on respiratory isolates, TOL/TAZ was effective against most Gram-negative pathogens (92% susceptible for *P. aeruginosa*, >80% for overall *Enterobacterales*, 73% for carbapenem-non-susceptible *P. aeruginosa* and 66% for carbapenem-susceptible, ESBL-producing *K. Pneumoniae*) in the Asia-Pacific region [7].

Among clinical isolates from intensive care units in Taiwan, the susceptibility rates of TOL/TAZ were 93% for *P. aeruginosa*, 88% for *E. coli*, and 80% for *K. pneumoniae* [8]. Furthermore, among *E. coli* and *K. pneumoniae* bloodstream isolates, hospital-acquired *K. pneumoniae* isolates were significantly less susceptible to TOL/TAZ than hospital-acquired *E. coli* isolates. Hospital-acquired and phenotypically ESBL-producing *K. pneumoniae* strains have become a heavy resistance burden in Taiwan [9]. However, these studies were limited to some bacterial species, sources, and specific units. Therefore, this study was aimed to conduct a nationwide survey for the susceptibilities of *P. aeruginosa* and *Enterobacterales* of more diverse species collected from various clinical sources (mainly intra-abdominal, respiratory, and urinary specimens) of patients at multiple medical centers in Taiwan.

3. Materials and Methods

An *in-vitro* study was conducted by International Health Management Associates, Inc. (IHMA) on behalf of Merck. In this study the antibacterial activity of TOL/TAZ and ten comparator compounds was evaluated against a recent collection of bacterial isolates from the IHMA stock cultures (2012-2015) collected from medical centers in Taiwan. Minimum inhibitory concentrations (MICs) of TOL/TAZ, cefepime, cefoperazone-sulbactam, ceftazidime, ciprofloxacin, colistin, imipenem, levofloxacin, meropenem, piperacillin/tazobactam and tigecycline were determined using the reference Clinical and Laboratory Standards Institute (CLSI) broth microdilution method [10].

Quality control and interpretation of results were performed in accordance with CLSI M100-S30 [11]. Quality control strains were included on each day of testing. The in vitro activity of tigecycline for *Enterobacterales* was interpreted according to the FDA criteria: Susceptible, ≤ 2 μg/mL; Intermediate, 4 μg/mL; and Resistant, ≥ 8 μg/mL [12].

The CLSI breakpoints (M100-S30) for *Enterobacterales* for TOL/TAZ are Susceptible, MIC ≤ 2/4 μg/mL; Intermediate, 4/4 μg/mL; and Resistant, ≥ 8/4 μg/mL. For imipenem and meropenem, the breakpoints are Susceptible, ≤ 1 μg/mL; Intermediate, 2 μg/mL; and Resistant, ≥ 4 μg/mL. For colistin, the breakpoints are Intermediate, ≤ 2 mg/L; and Resistant, ≥ 4 μg/mL. The CLSI breakpoints for *Pseudomonas aeruginosa* for TOL/TAZ are Susceptible, ≤ 4/4 μg/mL; Intermediate, 8/4 μg/mL; and Resistant, ≥ 16/4 μg/mL. For imipenem and meropenem, the breakpoints are Susceptible, ≤ 2 μg/mL; Intermediate, 4 μg/mL; and Resistant, ≥ 8 μg/mL. For colistin, the breakpoints are Intermediate, ≤ 2 μg/mL, and Resistant, ≥ 4 μg/mL (no more susceptible criterion for colistin is defined).

4. Results

A total of 599 clinical isolates were collected from various sites and specimens (Table 1), including intra-abdomen (*n* = 223), respiratory tract (*n* = 153), urine (*n* = 149), skin and soft tissue (*n* = 55), blood (*n* = 11) and others (*n* = 8). The species of isolates included *Enterobacterales* (*n* = 446), *Pseudomonas aeruginosa* (*n* = 149), *and Streptococcus spp.* (*n* = 4). The isolates were collected from the years in 2012 *(n* = 15), 2013 (*n* = 8), 2014 (*n* =427), and 2015 (*n* =149). Nine medical centers contributed the isolates, including 2 centers in northern Taiwan, 3 in central Taiwan, and 4 in southern Taiwan.

TOL/TAZ demonstrated good in vitro activity against *Streptococcus spp.* (*n* = 4, MIC range, ≤ 0.06 - 4 μg/mL; CLSI susceptible breakpoint, ≤ 8/4 μg/mL), although the isolates were rare in the current cohort of isolates (Table 2).

Table 1. Distribution of isolates by species enrolled in this study

Organism	Number
Enterobacterales	**446**
Citrobacter freundii	15
Citrobacter koseri	15
Enterobacter aerogenes	15
Enterobacter cloacae	60
Escherichia coli	125
Klebsiella oxytoca	30
Klebsiella pneumoniae	125
Morganella morganii	15
Proteus mirabilis	30
Serratia marcescens	16
Pseudomonas aeruginosa	**149**
***Streptococcus* spp.**	**4**
Streptococcus anginosus	2
Streptococcus constellatus	2
Total	**599**

Table 2. Susceptibiltiy, MIC50/90 and Range of MICs by species for all 11 antibiotics tested

Organism (no. of isolates)	Antimicrobial	%S	%I	%R	MIC_{50}	MIC_{90}	Range
Enterobacterales (446)	Ceftolozane/tazobactam	84.8	3.6	11.7	0.25	8	≤ 0.06 - > 32
	Cefepime	82.3	7.0	10.8	≤ 0.25	16	≤ 0.25 - > 32
	Cefoperazone/sulbactam	na	na	na	0.5	32	≤ 0.25 - > 64
	Ceftazidime	70.9	1.8	27.4	0.5	> 32	≤ 0.12 - > 32
	Ciprofloxacin	74.4	1.1	24.4	≤ 0.03	> 8	≤ 0.03 - > 8
	Colistin	0	84.1	15.9	0.5	> 8	≤ 0.12 - > 8
	Imipenem	83.2	9.0	7.9	0.5	2	≤ 0.06 - > 8
	Levofloxacin	75.1	2.2	22.7	0.12	> 8	≤ 0.06 - > 8
	Meropenem	97.8	0.7	1.6	≤ 0.12	≤ 0.12	≤ 0.12 - > 8
	Piperacillin/tazobactam	81.6	3.6	14.8	4	> 128	≤ 0.5 - > 128
	Tigecycline	88.6	7.9	3.6	1	4	0.12 - > 8
Citrobacter freundii (15)	Ceftolozane/tazobactam	60.0	20	20.0	2	32	≤ 0.06 - > 32
	Cefepime	86.7	0	13.3	0.5	32	≤ 0.25 - > 32
	Cefoperazone/sulbactam	na	na	na	16	32	≤ 0.25 - 64
	Ceftazidime	46.7	0	53.3	> 32	> 32	0.5 - > 32
	Ciprofloxacin	86.7	6.7	6.7	0.06	2	≤ 0.03 - 8
	Colistin	0	100	0	0.5	0.5	0.25 - 1
	Imipenem	73.3	20	6.7	1	2	0.5 - 4
	Levofloxacin	86.7	6.7	6.7	0.25	4	≤ 0.06 - 8
	Meropenem	100	0	0	≤ 0.12	≤ 0.12	≤ 0.12 - 1
	Piperacillin/tazobactam	46.7	26.7	26.7	32	128	1 - > 128
	Tigecycline	86.7	13.3	0	1	4	0.25 - 4

Organism (no. of isolates)	Antimicrobial	%S	%I	%R	MIC_{50}	MIC_{90}	Range
Citrobacter koseri (15)	Ceftolozane/tazobactam	93.3	0	6.7	0.12	0.25	≤ 0.06 - > 32
	Cefepime	93.3	0	6.7	≤ 0.25	≤ 0.25	≤ 0.25 - > 32
	Cefoperazone/sulbactam	na	na	na	≤ 0.25	≤ 0.25	≤ 0.25 - > 64
	Ceftazidime	93.3	0	6.7	0.25	0.5	0.25 - > 32
	Ciprofloxacin	100	0	0	≤ 0.03	≤ 0.03	≤ 0.03 - 0.25
	Colistin	0	100	0	0.25	0.5	0.25 - 0.5
	Imipenem	100	0	0	0.25	0.25	0.12 - 0.5
	Levofloxacin	100	0	0	≤ 0.06	0.12	≤ 0.06 - 0.25
	Meropenem	100	0	0	≤ 0.12	≤ 0.12	≤ 0.12 - ≤ 0.12
	Piperacillin/tazobactam	93.3	0	6.7	4	4	1 - > 128
	Tigecycline	100	0	0	0.5	1	0.12 - 1
Enterobacter cloacae (60)	Ceftolozane/tazobactam	76.7	5.0	18.3	0.5	16	0.12 - > 32
	Cefepime	88.3	8.3	3.3	≤ 0.25	4	≤ 0.25 - > 32
	Cefoperazone/sulbactam	na	na	na	≤ 0.25	32	≤ 0.25 - 32
	Ceftazidime	73.3	1.7	25.0	1	> 32	≤ 0.12 - > 32
	Ciprofloxacin	91.7	1.7	6.7	≤ 0.03	1	≤ 0.03 - > 8
	Colistin	0	86.7	13.3	0.5	> 8	0.25 - > 8
	Imipenem	80.0	16.7	3.3	1	2	0.12 - 4
	Levofloxacin	91.7	1.7	6.7	≤ 0.06	2	≤ 0.06 - > 8
	Meropenem	100	0	0	≤ 0.12	0.25	≤ 0.12 - 0.5
	Piperacillin/tazobactam	73.3	5.0	21.7	4	> 128	1 - > 128
	Tigecycline	93.3	3.3	3.3	1	2	0.12 - 8

Table 2. (Continued)

Organism (no. of isolates)	Antimicrobial	%S	%I	%R	MIC_{50}	MIC_{90}	Range
Escherichia coli (125)	Ceftolozane/tazobactam	93.6	1.6	4.8	0.25	1	$\leq$ 0.06 - > 32
	Cefepime	77.6	9.6	12.8	$\leq$ 0.25	16	$\leq$ 0.25 - > 32
	Cefoperazone/sulbactam	na	na	na	2	16	$\leq$ 0.25 - > 64
	Ceftazidime	65.6	4.0	30.4	0.5	32	$\leq$ 0.12 - > 32
	Ciprofloxacin	59.2	0.8	40.0	0.5	> 8	$\leq$ 0.03 - > 8
	Colistin	0	98.4	1.6	0.25	0.5	0.25 - 8
	Imipenem	100	0	0	0.25	0.5	$\leq$ 0.06 - 1
	Levofloxacin	59.2	0.8	40.0	0.5	> 8	$\leq$ 0.06 - > 8
	Meropenem	100	0	0	$\leq$ 0.12	$\leq$ 0.12	$\leq$ 0.12 - 0.5
	Piperacillin/tazobactam	93.6	3.2	3.2	4	8	$\leq$ 0.5 - > 128
	Tigecycline	100	0	0	0.5	1	0.12 - 2
Klebsiella aerogenes (15)	Ceftolozane/tazobactam	66.7	20.0	13.3	0.5	8	0.12 - 8
	Cefepime	93.3	6.7	0	$\leq$ 0.25	2	$\leq$ 0.25 - 4
	Cefoperazone/sulbactam	na	na	na	1	16	$\leq$ 0.25 - 32
	Ceftazidime	53.3	0	46.7	4	> 32	0.25 - > 32
	Ciprofloxacin	86.7	0	13.3	$\leq$ 0.03	4	$\leq$ 0.03 - 4
	Colistin	0	100	0	0.5	1	0.25 - 1
	Imipenem	60.0	33.3	6.7	1	2	0.12 - 4
	Levofloxacin	86.7	0	13.3	$\leq$ 0.06	8	$\leq$ 0.06 - > 8
	Meropenem	93.3	6.7	0	$\leq$ 0.12	1	$\leq$ 0.12 - 2
	Piperacillin/tazobactam	66.7	6.7	26.7	8	128	$\leq$ 0.5 - 128
	Tigecycline	80.0	20.0	0	1	4	0.25 - 4

Organism (no. of isolates)	Antimicrobial	%S	%I	%R	MIC_{50}	MIC_{90}	Range
Klebsiella pneumoniae (125)	Ceftolozane/tazobactam	75.2	2.4	22.4	0.25	> 32	≤ 0.06 - > 32
	Cefepime	72.8	7.2	20.0	≤ 0.25	> 32	≤ 0.25 - > 32
	Cefoperazone/sulbactam	na	na	na	≤ 0.25	> 64	≤ 0.25 - > 64
	Ceftazidime	64.0	1.6	34.4	0.5	> 32	≤ 0.12 - > 32
	Ciprofloxacin	67.2	0	32.8	0.12	> 8	≤ 0.03 - > 8
	Colistin	0	99.2	0.8	0.5	1	≤ 0.12 - > 8
	Imipenem	88.0	5.6	6.4	0.25	2	0.12 - > 8
	Levofloxacin	67.2	4.0	28.8	0.25	> 8	≤ 0.06 - > 8
	Meropenem	93.6	1.6	4.8	≤ 0.12	0.25	≤ 0.12 - > 8
	Piperacillin/tazobactam	67.2	3.2	29.6	4	> 128	≤ 0.5 - > 128
	Tigecycline	91.2	8.0	0.8	1	2	0.25 - 8
Morganella morgani (15)	Ceftolozane/tazobactam	100	0	0	0.25	0.5	0.12 - 1
	Cefepime	100	0	0	≤ 0.25	≤ 0.25	≤ 0.25 - ≤ 0.25
	Cefoperazone/sulbactam	na	na	na	0.5	4	≤ 0.25 - 4
	Ceftazidime	86.7	0	13.3	0.25	16	≤ 0.12 - 32
	Ciprofloxacin	86.7	0	13.3	≤ 0.03	4	≤ 0.03 - > 8
	Colistin	0	0	100	> 8	> 8	4 - > 8
	Imipenem	0	20.0	80.0	4	8	2 - 8
	Levofloxacin	86.7	0	13.3	≤ 0.06	8	≤ 0.06 - 8
	Meropenem	100	0	0	≤ 0.12	≤ 0.12	≤ 0.12 - ≤ 0.12
	Piperacillin/tazobactam	100	0	0	≤ 0.5	1	≤ 0.5 - 16
	Tigecycline	66.7	26.7	6.7	2	4	0.5 - 8

Table 2. (Continued)

Organism (no. of isolates)	**Antimicrobial**	**%S**	**%I**	**%R**	**MIC_{50}**	**MIC_{90}**	**Range**
Proteus mirabilis (30)	Ceftolozane/tazobactam	100	0	0	0.5	0.5	0.12 - 1
	Cefepime	96.7	3.3	0.0	≤ 0.25	≤ 0.25	≤ 0.25 - 8
	Cefoperazone/sulbactam	na	na	na	0.5	4	≤ 0.25 - 8
	Ceftazidime	96.7	0	3.3	≤ 0.12	≤ 0.12	≤ 0.12 - 32
	Ciprofloxacin	73.3	6.7	20.0	0.06	4	≤ 0.03 - > 8
	Colistin	0	0	100	> 8	> 8	> 8 - > 8
	Imipenem	33.3	30.0	36.7	2	4	0.12 - 8
	Levofloxacin	83.3	3.3	13.3	0.12	8	≤ 0.06 - > 8
	Meropenem	100	0	0	≤ 0.12	≤ 0.12	≤ 0.12 - 0.25
	Piperacillin/tazobactam	100	0	0	≤ 0.5	1	≤ 0.5 - 4
	Tigecycline	26.7	40.0	33.3	4	8	0.5 - 8
Serratia marcescens (16)	Ceftolozane/tazobactam	81.3	12.5	6.3	0.5	4	0.12 - 8
	Cefepime	81.3	6.3	12.5	≤ 0.25	16	≤ 0.25 - 32
	Cefoperazone/sulbactam	na	na	na	1	16	≤ 0.25 - 16
	Ceftazidime	87.5	0	12.5	0.25	> 32	0.25 - > 32
	Ciprofloxacin	81.3	0	18.8	0.12	8	≤ 0.03 - > 8
	Colistin	0	6.3	93.8	> 8	> 8	1 - > 8
	Imipenem	87.5	12.5	0	1	2	0.25 - 2
	Levofloxacin	81.3	6.3	12.5	0.25	8	0.12 - 8
	Meropenem	93.8	0	6.3	≤ 0.12	1	≤ 0.12 - > 8
	Piperacillin/tazobactam	100	0	0	2	8	≤ 0.5 - 16
	Tigecycline	75.0	12.5	12.5	2	> 8	1 - > 8

Organism (no. of isolates)	Antimicrobial	%S	%I	%R	MIC_{50}	MIC_{90}	Range
Pseudomonas aeruginosa (149)	Ceftolozane/tazobactam	94.0	3.4	2.7	0.5	4	0.12 - > 32
	Cefepime	63.8	22.2	14.1	8	32	0.5 - > 32
	Cefoperazone/sulbactam	na	na	na	8	64	1 - > 64
	Ceftazidime	61.8	6.7	31.5	4	> 32	1 - > 32
	Ciprofloxacin	67.1	4.0	28.9	0.12	> 8	≤ 0.03 - > 8
	Colistin	0	100	0	1	2	≤ 0.12 - 2
	Imipenem	59.1	10.1	30.9	2	> 8	1 - > 8
	Levofloxacin	64.4	4.0	31.5	0.5	> 8	≤ 0.06 - > 8
	Meropenem	70.5	3.4	26.2	0.5	> 8	≤ 0.12 - > 8
	Piperacillin/tazobactam	53.7	13.4	32.9	16	> 128	2 - > 128
	Tigecycline	na	na	na	8	> 8	2 - > 8
Streptococcus anginosus (2)	Ceftolozane/tazobactam	100	0	0	--	--	4 - 4
	Cefepime	100	0	0	--	--	≤ 0.25 - 0.5
	Cefoperazone/sulbactam	na	na	na	--	--	≤ 0.25 - ≤ 0.25
	Ceftazidime	na	na	na	--	--	1 - 2
	Ciprofloxacin	na	na	na	--	--	0.06 - 0.5
	Colistin	na	na	na	--	--	1 - > 8
	Imipenem	na	na	na	--	--	≤ 0.06 - ≤ 0.06
	Levofloxacin	100	0	0	--	--	≤ 0.06 - 0.5
	Meropenem	100	0	0	--	--	≤ 0.12 - ≤ 0.12
	Piperacillin/tazobactam	na	na	na	--	--	≤ 0.5 - ≤ 0.5
	Tigecycline	na	na	na	--	--	0.03 - 0.12

Table 2. (Continued)

Organism (no. of isolates)	Antimicrobial	%S	%I	%R	MIC_{50}	MIC_{90}	Range
Streptococcus constellatus (2)	Ceftolozane/tazobactam	100	0	0	--	--	≤ 0.06 - 2
	Cefepime	100	0	0	--	--	≤ 0.25 - ≤ 0.25
	Cefoperazone/sulbactam	na	na	na	--	--	≤ 0.25 - ≤ 0.25
	Ceftazidime	na	na	na	--	--	0.25 - 2
	Ciprofloxacin	na	na	na	--	--	≤ 0.03 - 0.12
	Colistin	na	na	na	--	--	> 8 - > 8
	Imipenem	na	na	na	--	--	≤ 0.06 - ≤ 0.06
	Levofloxacin	100	0	0	--	--	≤ 0.06 - ≤ 0.06
	Meropenem	100	0	0	--	--	≤ 0.12 - ≤ 0.12
	Piperacillin/tazobactam	na	na	na	--	--	≤ 0.5 - ≤ 0.5
	Tigecycline	na	na	na	--	--	≤ 0.015 - 0.06

%S, I, R: percent susceptible, intermediate, resistant according to CLSI 2016 (for ceftolozane/tazobactam, cefepime, ceftazidime, ciprofloxacin, imipenem, levofloxacin, meropenem, and piperacillin/tazobactam), EUCAST 2016 (for colistin), FDA (for tigecycline) guidelines; MIC_{50}, MIC_{90}, and range in µg/mL; na; no established breakpoints; --: MIC_{50}, MIC_{90} not calculated for only few isolates ($n = 2$).

TOL/TAZ demonstrated good *in vitro* activity against 84.8% of all *Enterobacterales* isolates with an MIC_{50} level of 0.25 μg/mL and an MIC_{90} of 8 μg/mL (Table 2). The overall activity of TOL/TAZ versus *Enterobacterales* was similar to those of ceftazidime (70.9% susceptible), PIP/TAZ (81.6% susceptible), cefepime (82.3% susceptible), imipenem (83.2% susceptible), and tigecycline (88.6% susceptible). The susceptibility rates of *K. pneumoniae* were TOL/TAZ (75.2%), cefepime (72.8%), PIP/TAZ (67.2%), and ceftazidime (64.0%). Additionally, the susceptibility rates of *E. coli* were TOL/TAZ (93.6%), cefepime (77.6%), and ceftazidime (65.6%). Tigecycline has better activity versus *K. pneumoniae* (91.2% susceptible) and *E. coli* (100% susceptible) in comparison to TOL/TAZ. However, tigecycline has poorer activity versus *Morganella morganii* (66.7% susceptible) and *Proteus mirabilis* (26.7% susceptible) than TOL/TAZ (100% active versus both species). Intrinsically, colistin was least active to *M. morganii* (100% resistant), *P. mirabilis* (100% resistant) and *Serratia marcescens* (93.8% resistant).

Of the ceftazidime non-susceptible *K. pneumoniae*, 31.1% of the isolates were susceptible to TOL/TAZ, with similar activity of TOL/TAZ versus those isolates from most common sources of respiratory, intra-abdominal and urinary sources (Table 3). For comparison, the activity of TOL/TAZ was similar to cefepime (31.1%), better than ciprofloxacin (6.7%) and piperacillin/tazobactam (17.8%), and inferior to meropenem (82.2%) and tigecycline (82.2%). Colistin remained the lowest MIC level (MIC_{90} = 1 μg/mL), but no susceptible crterion could be defined by CLSI 2020 (Table 3).

TOL/TAZ has diverse activities versus traditional AmpC producers. Firstly, *Citrobacter koseri* (*n* = 15, MIC_{90} = 0.25 μg/mL) and *M. morganii* (*n* = 15, MIC_{90} = 0.5 μg/mL) were highly susceptible to TOL/TAZ (93.3% and 100% susceptible, respectively). Secondly, *Enterobacter cloacae* (*n* = 60, MIC_{90} = 4 μg/mL) and *S. marcescens* (*n* = 16, MIC_{90} = 4 μg/mL) were moderately susceptible to TOL/TAZ (76.7% and 81.3% susceptible, respectively). Thirdly, *C. freundii* (*n* = 15, MIC_{90} = 32 μg/mL) and *K. aerogenes* (*n* = 15, MIC_{90} = 8 μg/mL) had the lowest susceptible rates to TOL/TAZ (60.0% and 66.7% susceptible, respectively). The third group has low susceptibility to ceftazidime (46.7% and 53.3%, respectively), but they remained highly susceptible to cefepime (86.7% and 96.3% susceptible, respectively), consistent with the AmpC phenotype (Table 2).

Table 3. Susceptibiltiy, $MIC_{50/90}$ and range of MICs by ceftazidime-nuncescetible *Klebsiella pneumoiae* and that from most common sources for antibiotics tested

Organism/source (no. of isolates)	**Antimicrobial**	**%S**	**%I**	**%R**	**MIC_{50}**	**MIC_{90}**	**Range**
Klebsiella pneumoniae non-susceptible to ceftazidime (45)	Ceftolozane/ tazobactam	31.1	6.7	62.2	32	>32	0.12 - >32
	Cefepime	31.1	17.8	51.1	16	>32	<0.25- >32
	Ciprofloxacin	6.7	4.4	88.9	>8	>8	<0.03 - >8
	Colisitn	0	97.8	2.2	0.5	1	0.25 - >8
	Meropenem	82.2	4.4	13.3	<0.12	8	<0.12 - >8
	Piperacillin/ tazobactam	17.8	6.7	75.5	>128	>128	2 - >128
	Tigecycline	82.2	15.6	2.2	1	4	0.5 - 8
Respiratory source (16)	Ceftolozane/ tazobactam	31.3	0	68.7	32	>32	0.5 - >32
Intra-abdominal source (15)	Ceftolozane/ tazobactam	33.3	13.3	53.3	16	>32	0.12 - >32
Urinary source (11)	Ceftolozane/ tazobactam	27.3	9.1	63.6	32	>32	2 - >32

With regard to *P. aeruginosa*, TOL/TAZ had the best activity (94% susceptibility, MIC_{90} = 4 μg/mL) in comparison to PIP/TAZ (54% susceptibility, MIC_{90} >128 μg/mL), ceftazidime (61.8% susceptibility, MIC_{90} >32 μg/mL), cefepime (64% susceptibility; MIC_{90} = 32 μg/mL), imipenem (59% susceptibility, MIC90 >8 μg/mL), and meropenem (71% susceptibility, MIC_{90} >8 μg/mL). According to CLSI 2020 breakpoints, all *P. aeruginosa* isolates were intermediate to colistin (Table 2).

5. Discussion

TOL/TAZ has higher activity versus *Enterobacterales* (84.8% susceptible) than ceftazidime (70.9% susceptible). Tazobactam may enhance or preserve the activity of ceftozolane by avoiding hydrolysis of some ESBLs or AmpC β-lactamases, but not carbapenemases [13]. The discrepancy of activity between TOL/TAZ and ceftazidime versus *K. pneumoniae* and *E. coli* of current data (75.2 vs. 64.0% for *K. pneumoniae* and 93.6 vs. 65.6% for *E. coli*) indicated different components of ESBLs, AmpCs or carbapenemases between the two *Enterobacterales* species. For example, carbapenemase production seems higher in *K. pneumoniae* isolates than in *E. coli* isolates, as TOL/TAZ was 75.2% and 93.6% susceptible, respectively. TOL/TAZ remained active (31.1% susceptible) versus some of the ceftazidime-non-susceptible isolates, some of which are believed to produce carbapenemases in addition to others with ESBL or AmpC β-lactamases. The activity of TOL/TAZ in Taiwan was similar to a previous report for the isolates from lower respiratory tract infections in the Asia-Pacific region during 2015-2016, showing the discrepancy in susceptibility rates between ESBL-producing *K. pneumoniae* and *E. coli* for TOL/TAZ (50% vs. 80%) [7]. Moreover, the susceptible rates of ESBL-producing *K. pneumoniae* to TOL/TAZ could be lowest to near 20% for isolates with bla_{SHV}-ESBL [7]. Similarly, in another study of bloodstream isolates from 16 major teaching hospitals in Taiwan in 2017, up to 80% of phenotypically ESBL-producing *K. pneumoniae* isolates were nonsusceptible to TOL/TAZ [9].

The traditional AmpC producers in *Enterobacterales* could be classified into 3 groups according to their susceptibilities to TOL/TAZ, based on different levels of a*mpC* gene expression. Besides, tazobactam is a well-known inducer of AmpC. Group 1 producers like *C. koseri* and *M. morganii* were highly susceptible to TOL/TAZ, indicating low-level expression of AmpC β-lactamase. Group 2 producers like *E. cloacae* and *S. marcescens* were moderately susceptible to TOL/TAZ, indicating inducible AmpC production by tazobactam. Group 3 producers like *C. freundii* and *K. aerogenes* were least susceptible to TOL/TAZ, indicating AmpC hyperproduction or de-repressive mutants [13].

Ceftozolane has intrinsically high activities against *P. aeruginosa*, even these isolates have heterogenous and multifactorial mechanisms conferring resistance to PIP/TAZ, ceftazidime, cefepime, imipenem and meropenem. Cefotozalane is capable to evade many of the resistance mechanisms employed by *P. aeruginosa*, including efflux pumps, reduced uptake due to porin defects, and modification of PBPs, but not the production of carbapenemases [1]. Carbapenemases might play a minor role among these TOL/TAZ-susceptible and carbapenem-resistant *P. aeruginosa* isolates, because neither ceftozolane nor tazobactam could resist carbapenemases. Among isolates collected from intensive care units of multiple medical centers in Taiwan, carbapenemase-encoding genes like bla_{KPC} and $bla_{OXA\text{-}48\text{-}like}$ genes were not detected in *P. aeruginosa* isolates that were not susceptible to imipenem [8]. Therefore, TOL/TAZ has often exhibited good activity against imipenem-resistant *P. aeruginosa* in Taiwan, as these isolates are often not carbapenemase producers.

One major resistance mechanism of carbapenem-resistant *P. aeruginosa* was outer membrane protein efflux pump gene (*oprD*) nucleotide substitution or deletion, which was highly prevalent (95.2%) among carbapenem-resistant *P. aeruginosa* isolates in Taiwan [14]. Decreased expression of OprD may confer carbapenem resistance to *P. aeruginosa*, especially to imipenem, but would result in less meropenem resistance [15-17], thus leading to the discrepancy of susceptibility between imipenem (59.1% susceptible) and meropenem (70.5% susceptible) in the current study. However, ceftozolane remains active against efflux pump-mediated carbapenem-resistant *P. aeruginosa*.

The clinical efficacy of piperacillin/tazobactam and cefepime on in vitro-susceptible ESBL-producing *Enterobacterales* remains controversial. Carbapenems are the most reliable antibiotics for severe infections caused by ESBL-producing *Enterobacterales*. However, their overuse has led to a serious problem of increasing drug resistance. TOL/TAZ offers new options for the treatment of ESBL infections, if not co-exisitng carbapenemases [18]. Meanwhile, the rise in resistant Gram-negative pathogens continues to challenge clinicians in treating serious infections. TOL/TAZ is a promising new agent for the treatment of complicated urinary tract infections (cUTIs), complicated intra-abdominal infections (cIAIs) in combination with metronidazole, and nosocomial/ventilator-associated pneumonia (VAP), including those caused by multidrug-resistant gram-negative organisms [1, 2, 18].

Phase III clinical trials for the treatment of cUTIs and cIAIs have been completed, showing that TOL/TAZ is non-inferior to meropenem and is an effective and a safe alternative for the treatment of these infections. The approved dose for cUTIs and cIAIs is 1.5 g (1 g ceftolozane and 500 mg tazobactam) infused over 1 hour every 8 hours [19, 20]. Recent data suggest that TOL/TAZ is a promising alternative agent for the treatment of cUTIs and cIAIs with adequate clinical cure rates, including those caused by ESBL-producing *Enterobacterales* and multidrug-resistant *P. aeruginosa* [21, 22].

TOL/TAZ was the most active β-lactam tested against *P. aeruginosa* isolates from from pneumonia and displayed *in vitro* activity against *Enterobacterales*, including isolates resistant to cephalosporins and carrying ESBL and *ampC* genes [5]. The phase III ASPECT-NP trial revealed that TOL/TAZ 3g (2 g ceftozolane and 1 g tazobactam) was well tolerated and was non-inferior to meropenem in terms of both 28-day all-cause mortality (24% vs. 25%) and clinical cure (54% vs. 53%) for the treatment of Gram-negative nosocomial pneumonia in mechanically ventilated critically ill patients [23].

With regard to antimicrobial stewardship to aid decision-making on the targeting of resources for antibiotic therapy, the empirical use of TOL/TAZ is cost-effective and improves quality-adjusted life years compared to piperacillin/tazobactam for the treatment of cIAIs and cUTIs at risk of resistant infection, by using different economic analytic models performed in the US, UK, or Taiwan [24-27].

Conclusion

Ceftolozane/tazobactam (TOL/TAZ) is active against a wide range of Gram-negative pathogens, including *Enterobacterales*, *P. aeruginosa*, and some streptococcal species, mainly isolated from intra-abdominal, respiratory and urinary specimens in the current study. TOL/TAZ was highly active versus cephalosporin- and carbapenem-resistant *P. aeruginosa*. TOL/TAZ also demonstrated higher *in-vitro* activity than currently available cephalosporins (ceftazidime and cefepime) and piperacillin/tazobactam when tested against various species of *Enterobacterales*.

Acknowledgments

The technical and molecular support for this study provided by IHMA, Inc. is gratefully acknowledged. We thank Mei-Yu Su at Department of Intensive Care Medicine, Chi Mei Medical Center, for data collection.

References

[1] Zhanel GG, Chung P, Adam H, Zelenitsky S, Denisuik A, et al. Ceftolozane/tazobactam: a novel cephalosporin/β-lactamase inhibitor combination with activity against multidrug-resistant gram-negative bacilli. *Drugs*, 2014, 74, 31-51.

[2] Cho JC, Fiorenza MA, Estrada SJ. Ceftolozane/Tazobactam: A novel cephalosporin/β-lactamase inhibitor combination. *Pharmacotherapy*, 2015, 35, 701-715

[3] Pfaller MA, Bassetti M, Duncan LR, Castanheira M. Ceftolozane/tazobactam activity against drug-resistant *Enterobacteriaceae* and *Pseudomonas aeruginosa* causing urinary tract and intraabdominal infections in Europe: report from an antimicrobial surveillance programme (2012-15). *J Antimicrob Chemother.*, 2017, 72, 1386-1395.

[4] Shortridge D, Castanheira M, Pfaller MA, Flamm RK. Ceftolozane-tazobactam activity against *Pseudomonas aeruginosa* clinical isolates from U.S. hospitals: Report from the PACTS Antimicrobial Surveillance Program, 2012 to 2015. *Antimicrob Agents Chemother.*, 2017, 61, e00465-17.

[5] Castanheira M, Duncan LR, Mendes RE, Sader HS, Shortridge D. Activity of ceftolozane-tazobactam against *Pseudomonas aeruginosa* and *Enterobacteriaceae* isolates collected from respiratory tract specimens of hospitalized patients in the United States during 2013 to 2015. *Antimicrob Agents Chemother.*, 2018, 62, e02125-17

[6] Pfaller MA, Shortridge D, Sader HS, Castanheira M, Flamm RK. Ceftolozane/tazobactam activity against drug-resistant *Enterobacteriaceae* and *Pseudomonas aeruginosa* causing healthcare-associated infections in the Asia-Pacific region (minus China, Australia and New Zealand): report from an Antimicrobial Surveillance Programme (2013-2015). *Int J Antimicrob Agents*, 2018, 51, 181-189.

[7] Kuo SC, Liu CE, Lu PL, Chen YS, Lu MC, et al. Activity of ceftolozane-tazobactam against Gram-negative pathogens isolated from lower respiratory tract infections in the Asia-Pacific region: SMART 2015-2016. *Int J Antimicrob Agents*, 2020, 55, 105883.

[8] Liao CH, Lee NY, Tang HJ, Lee SS, Lin CF, et al. Antimicrobial activities of ceftazidime-avibactam, ceftolozane-tazobactam, and other agents against *Escherichia coli*, *Klebsiella pneumoniae*, and *Pseudomonas aeruginosa* isolated

from intensive care units in Taiwan: results from the Surveillance of Multicenter Antimicrobial Resistance in Taiwan in 2016. *Infect Drug Resist.*, 2019, 12, 545-552.

[9] Jean SS, Lu MC, Shi ZY, Tseng SH, Wu TS, et al. In vitro activity of ceftazidime-avibactam, ceftolozane-tazobactam, and other comparable agents against clinically important Gram-negative bacilli: results from the 2017 Surveillance of Multicenter Antimicrobial Resistance in Taiwan (SMART). *Infect Drug Resist.*, 2018, 11, 1983-1992.

[10] Clinical and Laboratory Standards Institute. M07-A10. Methods for dilution antimicrobial susceptibility tests for bacteria that grow aerobically; approved standard – tenth edition, *CLSI*, 2015, Wayne, PA, USA

[11] Clinical and Laboratory Standards Institute. M100-S30. Performance standards for antimicrobial susceptibility testing: 30th informational supplement, *CLSI*, 2020, Wayne, PA, USA.

[12] FDA insert package https://www.fda.gov/drugs/development-resources/antibacterial-susceptibility-test-interpretive-criteria

[13] Ku YH, Yu WL. Cefoperazone/sulbactam: New composites against multiresistant gram negative bacteria? *Infect Genet Evol.*, 2021, 88, 104707.

[14] Lai CC, Chen CC, Lu YC, Chuang YC, Tang HJ. In vitro activity of cefoperazone and cefoperazone-sulbactam against carbapenem-resistant Acinetobacter baumannii and *Pseudomonas aeruginosa. Infect Drug Resist.*, 2018, 12, 25-29.

[15] Li S, Jia X, Li C, Zou H, Liu H, Guo Y, Zhang L. Carbapenem-resistant and cephalosporin-susceptible Pseudomonas aeruginosa: a notable phenotype in patients with bacteremia. *Infect Drug Resist.*, 2018, 11, 1225-1235.

[16] Khalili Y, Yekani M, Goli HR, Memar MY. Characterization of carbapenem-resistant but cephalosporin-susceptible Pseudomonas aeruginosa. *Acta Microbiol Immunol Hung.*, 2019, 66, 529-540.

[17] Li H, Luo YF, Williams BJ, Blackwell TS, Xie CM. Structure and function of OprD protein in Pseudomonas aeruginosa: from antibiotic resistance to novel therapies. *Int J Med Microbiol.*, 2012, 302, 63-68.

[18] Sheu CC, Lin SY, Chang YT, Lee CY, Chen YH, Hsueh PR. Management of infections caused by extended-spectrum β-lactamase-producing *Enterobacteriaceae*: current evidence and future prospects. *Expert Rev Anti Infect Ther.*, 2018, 16, 205-218.

[19] Giancola SE, Mahoney MV, Bias TE, Hirsch EB. Critical evaluation of ceftolozane-tazobactam for complicated urinary tract and intra-abdominal infections. *Ther Clin Risk Manag.*, 2016, 12, 787-797.

[20] Solomkin J, Hershberger E, Miller B, Popejoy M, Friedland I, et al. Ceftolozane/tazobactam plus metronidazole for complicated intra-abdominal infections in an era of multidrug resistance: Results from a randomized, double-blind, phase 3 trial (ASPECT-cIAI). *Clin Infect Dis.*, 2015, 60, 1462-1471.

[21] Lin SY, Huang CH, Ko WC, Chen YH, Hsueh PR. Recent developments in antibiotic agents for the treatment of complicated intra-abdominal infections. *Expert Opin Pharmacother.*, 2016, 17, 339-354.

[22] Syue LS, Chen YH, Ko WC, Hsueh PR. New drugs for the treatment of complicated intra-abdominal infections in the era of increasing antimicrobial resistance. *Int J Antimicrob Agents*., 2016, 47, 250-258.

[23] Kollef MH, Nováček M, Kivistik Ü, Réa-Neto Á, Shime N, et al. Ceftolozane-tazobactam versus meropenem for treatment of nosocomial pneumonia (ASPECT-NP): a randomised, controlled, double-blind, phase 3, non-inferiority trial. *Lancet Infect Dis*., 2019, 19, 1299-1311.

[24] Prabhu VS, Solomkin JS, Medic G, Foo J, Borse RH, et al. Cost-effectiveness of ceftolozane/tazobactam plus metronidazole versus piperacillin/tazobactam as initial empiric therapy for the treatment of complicated intra-abdominal infections based on pathogen distributions drawn from national surveillance data in the United States. *Antimicrob Resist Infect Control*., 2017, 6, 107.

[25] Prabhu V, Foo J, Ahir H, Sarpong E, Merchant S. Cost-effectiveness of ceftolozane/tazobactam plus metronidazole compared with piperacillin/tazobactam as empiric therapy for the treatment of complicated intra-abdominal infections based on the in-vitro surveillance of bacterial isolates in the UK. *J Med Econ*., 2017, 20, 840-849.

[26] Kauf TL, Prabhu VS, Medic G, Borse RH, Miller B, et al. Cost-effectiveness of ceftolozane/tazobactam compared with piperacillin/tazobactam as empiric therapy based on the *in-vitro* surveillance of bacterial isolates in the United States for the treatment of complicated urinary tract infections. *BMC Infect Dis*., 2017, 17, 314.

[27] Chen GJ, Pan SC, Foo J, Morel C, Chen WT, Wang JT. Comparing ceftolozane/tazobactam versus piperacillin/tazobactam as empiric therapy for complicated urinary tract infection in Taiwan: A cost-utility model focusing on gram-negative bacteria. *J Microbiol Immunol Infect*., 2019, 52, 807-815.

Chapter 7

Elevated Levels of Caspase-3 in Cerebrospinal Fluid of Newborns with Perinatal Asphyxia: A Biomarker of Failing Mitochondrial Bioenergetics

Puneet K. Samaiya[1,3], Ashok Kumar[2] and Sairam Krishnamurthy[1,*]

[1]Neurotherapeutics Laboratory, Pharmaceutical Engineering and Technology, Indian Institute of Technology (Banaras Hindu University), Varanasi, India

[2]Department of Pediatrics, Institute of Medical Sciences, Banaras Hindu University, Varanasi, UP, India

[3]Shri GS Institute of Technology and Science, Indore (MP), India

Abstract

Mitochondria act as metabolic hubs within the cell, which are essential for energy production to fulfill the bioenergetic and biosynthetic demands of the developing brain. Failing mitochondrial bioenergetics can play a central role in neuronal cell death in pathological conditions like perinatal asphyxia (PA), which can be associated with insult progression over days and the development of a more severe condition hypoxic-ischemic encephalopathy (HIE) in the neonates. Therefore, to evaluate the role of mitochondrial bioenergetics in PA/HIE, we estimated apoptotic markers like caspase-3 in cerebrospinal fluid (CSF) of newborns with asphyxia (n = 42) and control (n =17) at birth. Apgar scores < 7 at 5 min were considered as asphyxic. Further, through sarnat scoring, the asphyxic neonates were divided into different stages of HIE. CSF sample was collected as sample-1 on day-1 and sample-2 on day-7

* Corresponding Author's Email: saibliss@hotmail.com, ksairam.phe@iitbhu.ac.in.

In: Advances in Biology. Volume 1
Editor: Charles D. Grant
ISBN: 979-8-88697-201-6

after asphyxia. We observed a significant ($p < 0.05$) increase in the levels of caspase-3 in the HIE stage (s)-III compared to s-II and s-I and a marked increase in the control group in sample-1. Further, the levels of caspase-3 were markedly elevated in HIE s-III compared to s-II and s-I in sample-2, which depicts the persistence of neuronal insult. The increased levels of caspase-3 in the CSF of neonates with different stages of HIE after asphyxia indicate activation of apoptosis in the developing brain of neonates, which can lead to neurological abnormalities in the later stage of life.

Keywords: perinatal asphyxia (PA), cerebrospinal fluid (CSF), hypoxic-ischemic encephalopathy (HIE), caspase-3, apoptosis

Introduction

Perinatal asphyxia (PA) in progression over days to Hypoxic–ischemic encephalopathy (HIE) can result in permanent neurological and cognitive impairment in children on a long-term basis (Volpe 2008). It is estimated that PA causes around 23% of all newborn deaths worldwide. India is the tenth country that accounts for more than 65% of all intrapartum-related perinatal deaths (deaths in the first week of life) imparted a result of PA (Lawn et al., 2010). The pathogenesis of cell death after a PA/hypoxic-ischemic encephalopathy (HIE) to the developing brain is complex and still not completely understood.

However, it has been stated that HIE followed by resuscitation appears to progress in two distinct phases, a primary cellular injury during the period of the insult due to necrosis and later to a delayed secondary injury 8-48 hours after the initial insult by apoptosis (Riikonen, Korhonen, and Lindholm 1999). This delayed or secondary energy failure during the recovery period after the birth injury is due to lack of perfusion, energy failure, glutamate excitotoxicity, and subsequent reperfusion, which is the ultimate cause of brain injury damage (Samaiya, Krishnamurthy, and Kumar 2021).

Analytical Framework

Literature Review

Primary neuronal injury (within minutes) in PA is the outcome of improper mitochondrial functioning and results mainly in necrotic cell death (Lai and

Yang 2010, Samaiya et al., 2016, Edwards et al., 1997). After several days to months, secondary neuronal injury and tertiary neuronal injury from months to years occur in apoptotic cell death, resulting in the spreading of the damage to adjacent, otherwise uninjured tissues (Fleiss and Gressens 2012). The mechanism of secondary neuronal injury results from several factors, including ischemia-reperfusion injury, oxidative stress, inflammation, rebound glutamate excitotoxicity, sequestration of mitochondrial calcium, and ultimately activation of proteases and caspases (Fleiss and Gressens 2012). Therefore, secondary damage is more detrimental than the initial insult (Park, Velumian, and Fehlings 2004, Dell'Anna et al., 1997). Hence, inhibition of secondary injury processes may be one of the essential means of therapeutic intervention to prevent further damage to neurons in the central nervous system (CNS) (Samaiya et al., 2016, 2017, Fleiss and Gressens 2012).

A previous study has reported that after hypoxia-ischemia, programmed cell death mechanisms are activated, and a considerable proportion of cells die not by necrosis but by apoptosis, in which caspases play a principal role (Yue et al., 1997). Caspases are a group of cysteine proteases essential for initiating and executing apoptosis. The caspase activation cascade takes place via two broadly recognized pathways: extrinsic and intrinsic. The extrinsic apoptotic pathway stimulates the death receptors (Fas/Apo1 or TRAIL) while the intrinsic apoptotic pathway involves the mitochondria, releasing cytochrome c from the mitochondrial intermembrane space cytosol (Nguyen, Gillet, and Popgeorgiev 2021, Rodriguez et al., 2021). Cytochrome c, an integral component of the mitochondrial electron transport chain, participates in ATP synthesis (Thornton and Hagberg 2015). Released cytochrome c in the cytosol results in activation of factor (Apaf-1) and caspase-9, leading to the formation of a high-molecular-mass cytoplasmic complex referred to as apoptosome, which later activates Caspase-3 and finally results in apoptotic cell death causing neurological defects later in life after PA (Chowdhury, Tharakan, and Bhat 2008, Kuranaga and Miura 2007, Szydlowska and Tymianski 2010).

Information Withholding Hypothesis

Thus, the present study hypothesizes that there may be a PA-induced release of cytochrome c in newborns' cerebrospinal fluid (CSF), which can initiate mitochondrial dysfunction and ultimately lead to neuronal cell death activating caspase-3.

Method

Patients

This prospective observational study was conducted at the Department of Pediatrics, Institute of Medical Sciences and Department of Pharmaceutical engineering and technology, Indian Institute of Technology, Banaras Hindu University. Informed consent was taken from all parents before inclusion in the study. The Institute Ethics Committee approved the study (No. Dean/2012-13/543). The group was divided into birth asphyxia (n = 42) and controls (n = 17).

Inclusion Criteria

The study population comprised full-term newborns with a gestational age ≥ 37 weeks that were included in the study group based on the following criteria for asphyxia:

- Evidence of fetal distress
- Umbilical artery pH < 7
- Apgar scored <7 in the 5th minute
- Postnatal resuscitation with positive pressure ventilation.

The asphyxia group was further divided into different stages of HIE (s-I, s-II, and s-III) by sarnat and sarnat scoring (Sarnat and Sarnat 1976).

Exclusion Criteria

Antenatal babies with sepsis, respiratory distress syndrome, metabolic disorders, and major congenital malformations were excluded from the study.

CSF Sampling

Approximately 1.5 ml of CSF samples were collected two times by spinal taps. Firstly, within 24 hours of birth in intramural babies and on the day of admission in extramural infants, and second, seven days after the first sample was collected. Caspase-3 was estimated in the CSF collected and spun at 4000 rpm for 10 min in a centrifuge. Cell counts and CSF proteins were analyzed as routine clinical samples. Samples were stored at - 80°C until analysis.

Caspase-3 Measurements

Levels of human activated caspase-3 were measured in 59 CSF samples using enzyme-linked immunosorbent assays (ab168541; Abcam Plc., Cambridge, MA, USA) according to the manufacturer's instructions. The detection level for caspase-3 is 2 μg/ml.

Statistical Analysis

Data are expressed as means ± SD. Groups were compared using the Kruskal – Wallis or the Mann – Whitney U test. A $p < 0.05$ was considered significant.

Results and Discussion

In the present study, forty-two patients with asphyxia (HIE) and seventeen in control groups were included, of which eight patients had s-I HIE, six s-II HIE, and twenty-eight with s-III HIE. Both patient and control groups had a male prevalence. Gestational age and birth weight were similar in the four groups (controls, s-I HIE, s-II HIE, and s-III HIE) (Table 1). The incidence of perinatal asphyxia was higher in unbooked mothers, extramural deliveries, and those born through spontaneous vaginal delivery (SVD) compared to the emergency caesarian section. The incidence of fetal distress was higher in the HIE group than in the control.

Moreover, fetal distress was more in HIE s-III. However, there were no changes in gravidity, parity, gestation, and delivery through the meconium-stained amniotic fluid (MSAF) between the control and HIE groups. All patients of s-II and s-III HIE developed seizures, thirteen of twenty-eight s-III HIE faced shock, two of the eight patients in the s-I HIE, two of the six of s-II HIE, and twenty-one of twenty-eight s-III HIE group had an acute renal injury, five of stage I, one of s-II HIE and twenty-four of s-III HIE had respiratory distress. Two of twenty-eight s-III HIE had persistent pulmonary hypertension. Meanwhile, nineteen of s-III HIE expired shortly after birth.

Figure 1 depicts that the concentrations of caspase-3 were increased in infants with s-III HIE when compared to control ($^{a}p < 0.05$), s-I HIE ($^{b}p < 0.05$), and s-II HIE ($^{c}p < 0.05$) in sample-1. Within the asphyxia group, infants with s-III HIE had significantly increased concentrations of caspase-3 when compared to s-I ($^{b}p < 0.05$) and s-II ($^{c}p < 0.05$) HIE groups in sample-2.

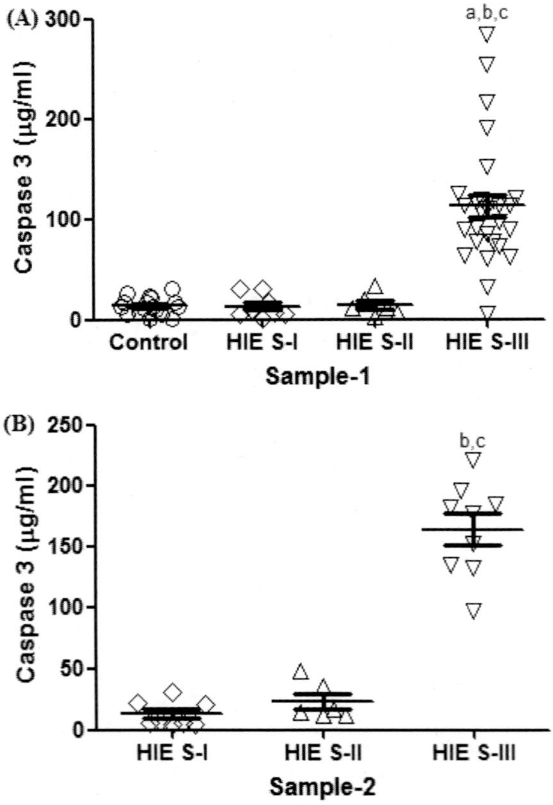

Figure 1. Levels of caspase-3 in CSF of control and different stages of HIE in (A) sampe-1 and comparison of the concentration of caspase-3 within HIE group (B) sample-2. All values are mean ± SD. ap < 0.05 compared to control, bp < 0.05 compared to Stage-I HIE and cp < 0.05 compared to Stage-II HIE groups [Kruskal – Wallis or the Mann – Whitney U test].

Table 1. Gestational age and birth weight of controls and patients (mean ± SD)

	Control (n = 17)	**Asphyxia (n = 42)**		
		HIE s-I (n = 8)	**HIE s-II (n = 6)**	**HIE s-III (n = 28)**
Birth weight(g)	2830 ± 435	2800 ± 460	2883 ± 365	2788 ± 342
Gestational age (Weeks)	38.2 ± 1.1	37.5 ± 1.1	37.8 ± 1.3	37.8 ± 1.2

This is the first study that suggests the increased levels of caspase-3 in CSF of neonates following neonatal anoxia. The levels of caspase-3 were more prominent in infants who later developed encephalopathy following anoxia than in the control group. Neuronal cell death post HIE is thought to be comprised of primary cell death (necrosis), caused by mechanical destruction, and subsequent or delayed cell death (apoptosis), caused by several pathologic cascades initiated by trauma (Eisenberg-Lerner et al., 2009). However, the mechanisms that lead to irreversible brain injury and apoptosis in HIE are complex and still partly unknown. In humans, these cascades include ischemia, excitotoxicity, and inflammation, each of which can trigger apoptosis (Satchell et al., 2005). Apoptosis is a tightly regulated homeostatic process necessary in embryologic development, and dysregulation of apoptosis can contribute to many human disease processes (Thompson 1995). Apoptosis has been considered the more prominent type of cell death in pediatric and adult patients (Northington, Chavez-Valdez, and Martin 2011). We have recently reported the involvement of mitochondrial dysfunction-linked apoptosis through cytochrome-C and caspase (caspase-9/3) in insult progression after anoxic injury (Samaiya et al., 2016). However, several other neural and biochemical markers in CSF have been reported to play an essential role in birth asphyxia (Ergenekon et al., 2004, Sävman et al., 2013).

Further, previous reports have suggested that secondary brain injury takes place due to impairment in mitochondrial respiration, membrane permeability transition, accumulation of calcium, changes in the Bcl-2 family proteins, the release of proapoptotic proteins (cytochrome c, apoptosis-inducing factor), and downstream activation of caspase-9 and caspase-3 occurs after hypoxia-ischemia in rats (Hagberg et al., 2009, Northington et al., 2007). Interestingly, we observed a marked increase in the levels of caspase-3 in sample-1 and sample-2 collected on d-1 and d-7, respectively, post-asphyxic injury. The levels of caspase-3 were significantly higher in the s-III HIE than s-I and s-II HIE groups, respectively depicting more severity of insult post-anoxia in s-III

HIE. A previous study defines the involvement of reactive oxygen species (ROS) in actuating these markers, which causes mitochondrial-linked apoptosis after an anoxic injury during HIE (Kim and Park 2003). These observations suggest that regulating apoptosis in the developing brain may be an important therapeutic target for neuroprotection after PA. Currently, neuroprotective interventions are limited to the first six hours of age.

Conclusion

This study attempts to explain the relationship between failing mitochondrial bioenergetics and activation of caspase-3 as a putative cell death pathway in perinatal asphyxia and identify clinical features that may provide insight into potential therapeutic targets patient populations.

Acknowledgment

SK is thankful to the Department of Biotechnology (DBT), New Delhi, India, for assistance in research grant [102/IFD/SAN/4654/2011-2012].

References

Chowdhury, I., Binu Tharakan, and Ganapathy K Bhat. 2008. "Caspases - an update." *Comparative Biochemistry and Physiology Part B: Biochemistry and Molecular Biology* 151 (1):10-27.

Dell'Anna, E., Y. Chen, E. Engidawork, K. Andersson, G. Lubec, J. Luthman, and M. Herrera-Marschitz. 1997. "Delayed neuronal death following perinatal asphyxia in rat." *Exp Brain Res* 115 (1):105-15.

Edwards, A. D., X. Yue, P. Cox, P. L. Hope, D. V. Azzopardi, M. V. Squier, and H. Mehmet. 1997. "Apoptosis in the brains of infants suffering intrauterine cerebral injury." *Pediatr Res* 42 (5):684-9. doi: https://10.1203/00006450-199711000-00022.

Eisenberg-Lerner, A., S. Bialik, H. U. Simon, and A. Kimchi. 2009. "Life and death partners: apoptosis, autophagy and the cross-talk between them." *Cell Death Differ* 16 (7):966-75. doi: https://10.1038/cdd.2009.33.

Ergenekon, E., K. Gücüyener, D. Erbaş, S. Aral, E. Koç, and Y. Atalay. 2004. "Cerebrospinal fluid and serum vascular endothelial growth factor and nitric oxide levels in newborns with hypoxic ischemic encephalopathy." *Brain Dev* 26 (5):283-6. doi: https://10.1016/s0387-7604(03)00166-9.

Fleiss, B., and P. Gressens. 2012. "Tertiary mechanisms of brain damage: a new hope for treatment of cerebral palsy?" *Lancet Neurol* 11 (6):556-66. doi: https://10.1016/s 1474-4422(12)70058-3.

Hagberg, H., C. Mallard, C. I. Rousset, and Wang Xiaoyang. 2009. "Apoptotic mechanisms in the immature brain: involvement of mitochondria." *J Child Neurol* 24 (9):1141-6. doi: https://10.1177/0883073809338212.

Kim, J. Y., and J. H. Park. 2003. "ROS-dependent caspase-9 activation in hypoxic cell death." *FEBS Lett* 549 (1-3):94-8. doi: https://10.1016/s0014-5793(03)00795-6.

Kuranaga, E., and Masayuki Miura. 2007. "Nonapoptotic functions of caspases: caspases as regulatory molecules for immunity and cell-fate determination." *Trends in cell biology* 17 (3):135-144.

Lai, M. C., and San-Nan Yang. 2010. "Perinatal hypoxic-ischemic encephalopathy." *BioMed Research International* 2011.

Lawn, Joy E, Kate Kerber, Christabel Enweronu-Laryea, and Simon Cousens. 2010. "3.6 million neonatal deaths - what is progressing and what is not?" *Seminars in perinatology*.

Nguyen, T. T. M., G. Gillet, and N. Popgeorgiev. 2021. "Caspases in the Developing Central Nervous System: Apoptosis and Beyond." *Front Cell Dev Biol* 9:702404. doi: https://10.3389/fcell.2021.702404.

Northington, F. J, Raul Chavez-Valdez, and Lee J Martin. 2011. "Neuronal cell death in neonatal hypoxia-ischemia." *Annals of neurology* 69 (5):743-758.

Northington, F. J., M. E. Zelaya, D. P. O'Riordan, K. Blomgren, D. L. Flock, H. Hagberg, D. M. Ferriero, and L. J. Martin. 2007. "Failure to complete apoptosis following neonatal hypoxia-ischemia manifests as "continuum" phenotype of cell death and occurs with multiple manifestations of mitochondrial dysfunction in rodent forebrain." *Neuroscience* 149 (4):822-33. doi: https://10.1016/j.neuroscience.2007. 06.060.

Park, E., A. A. Velumian, and M. G. Fehlings. 2004. "The role of excitotoxicity in secondary mechanisms of spinal cord injury: a review with an emphasis on the implications for white matter degeneration." *J Neurotrauma* 21 (6):754-74. doi: https://10.1089/0897715041269641.

Riikonen, R. S, Laura T Korhonen, and Dan B Lindholm. 1999. "Cerebrospinal nerve growth factor - a marker of asphyxia?" *Pediatric neurology* 20 (2):137-141.

Rodriguez, J., T. Li, Y. Xu, Y. Sun, and C. Zhu. 2021. "Role of apoptosis-inducing factor in perinatal hypoxic-ischemic brain injury." *Neural Regen Res* 16 (2):205-213. doi: https://10.4103/1673-5374.290875.

Samaiya, P. K., Gopeshwar Narayan, Ashok Kumar, and Sairam Krishnamurthy. 2016. "Neonatal anoxia leads to time dependent progression of mitochondrial linked apoptosis in rat cortex and associated long term sensorimotor deficits." *International Journal of Developmental Neuroscience* 52:55-65. doi: http://dx.doi.org/10.1016/j. ijdevneu.2016.05.005.

Samaiya, P. K., Gopeshwar Narayan, Ashok Kumar, and Sairam Krishnamurthy. 2017. "Tempol (4 hydroxy-tempo) inhibits anoxia-induced progression of mitochondrial dysfunction and associated neurobehavioral impairment in neonatal rats." *Journal of the Neurological Sciences* 375:58-67. doi: http://dx.doi.org/10.1016/j.jns.2017.01.021.

Samaiya, P. K., S. Krishnamurthy, and A. Kumar. 2021. "Mitochondrial dysfunction in perinatal asphyxia: role in pathogenesis and potential therapeutic interventions." *Mol Cell Biochem*. doi: https://10.1007/s11010-021-04253-8.

Sarnat, H. B., and M. S. Sarnat. 1976. "Neonatal encephalopathy following fetal distress. A clinical and electroencephalographic study." *Arch Neurol* 33 (10):696-705. doi: https://10.1001/archneur.1976.00500100030012.

Satchell, M. A., Y. Lai, P. M. Kochanek, S. R. Wisniewski, E. L. Fink, N. A. Siedberg, R. P. Berger, S. T. DeKosky, P. D. Adelson, and R. S. Clark. 2005. "Cytochrome c, a biomarker of apoptosis, is increased in cerebrospinal fluid from infants with inflicted brain injury from child abuse." *J Cereb Blood Flow Metab* 25 (7):919-27. doi: https://10.1038/sj.jcbfm.9600088.

Sävman, K., M. P. Heyes, P. Svedin, and A. Karlsson. 2013. "Microglia/macrophage-derived inflammatory mediators galectin-3 and quinolinic acid are elevated in cerebrospinal fluid from newborn infants after birth asphyxia." *Transl Stroke Res* 4 (2):228-35. doi: https://10.1007/s12975-012-0216-3.

Szydlowska, K., and Michael Tymianski. 2010. "Calcium, ischemia and excitotoxicity." *Cell calcium* 47 (2):122-129.

Thompson, C. B. 1995. "Apoptosis in the pathogenesis and treatment of disease." *Science* 267 (5203):1456-62. doi: https://10.1126/science.7878464.

Thornton, C., and H. Hagberg. 2015. "Role of mitochondria in apoptotic and necroptotic cell death in the developing brain." *Clin Chim Acta* 451 (Pt A):35-8. doi: https://10.1016/j.cca.2015.01.026.

Volpe, J. J. 2008. "*Neurology of the newborn E-book*."

Yue, X., H. Mehmet, J. Penrice, C. Cooper, E. Cady, J. S. Wyatt, E. O. R. Reynolds, A. D. Edwards, and M. V. Squier. 1997. "Apoptosis and necrosis in the newborn piglet brain following transient cerebral hypoxia-ischaemia." *Neuropathology and applied neurobiology* 23 (1):16-25.

Index

A

acute respiratory distress syndrome, ix, 124, 127, 132
additive properties, 156, 163
adenosine triphosphate, 82, 88, 91
antibiotic, ix, 19, 124, 128, 133, 134, 135, 138, 140, 201, 203
apoptosis, xi, 51, 71, 81, 129, 144, 206, 207, 211, 212, 213, 214
ATP-sensitive potassium channels (mKATP channels), vii, 49, 50, 51, 52, 53, 54, 55, 56, 57, 58, 59, 60, 61, 62, 63, 64, 65, 66, 67, 69, 70, 71, 72, 73, 74, 75, 76, 77, 78, 80, 82, 83

B

Bacillus subtilis, 35, 38, 40, 42, 43, 46
bacteremia, 125, 128, 132, 143, 203
bacteria, 9, 10, 16, 18, 25, 39, 42, 125, 129, 130, 134, 135, 147, 154, 203, 204
bacteriostatic, ix, 124, 133, 134
balanced biochemical equations, 153
bilirubin, vii, viii, 85, 86, 87, 88, 89, 90, 91, 92, 93, 94, 95, 96, 97, 98, 99, 100, 101, 102, 103, 104, 105, 106, 107, 108, 110, 111, 112, 113, 114, 115, 116, 117, 118, 119, 120, 121
biliverdin, 86, 105, 106, 115, 119, 120
biochemical reference state, x, 153, 155, 158
bioenergetics, vii, viii, xi, 49, 50, 51, 54, 60, 65, 66, 74, 76, 77, 83, 182, 205, 212
biological reference state (BRS), 154, 155, 158, 159, 167
biosynthesis, 34, 35, 40, 45, 115, 186
biotransformation, 1, 2, 4, 9, 13, 15, 17, 18, 19, 21, 28, 29, 39, 45
bronchial epithelium, 128, 130

C

Ca^{2+}, viii, 49, 51, 58, 65, 67, 68, 69, 71, 72, 73, 77, 169
cardiovascular diseases (CVD), 92
caspase-3, vii, xi, 129, 205, 206, 207, 209, 210, 211, 212
ceftolozane, 186, 196, 201, 202, 203, 204
ceftolozane/tazobactam (TOL/TAZ), v, x, xi, 185, 186, 187, 188, 189, 196, 197, 199, 200, 201, 202, 204
cerebrospinal fluid (CSF), vi, vii, xi, 104, 105, 119, 205, 206, 207, 208, 209, 210, 211, 214
chemical reference state (CRS), 158, 159, 160, 164, 167, 168, 169, 170
clindamycin, ix, 124, 133, 134, 140
Crigler-Najjar syndrome, 113
cytoplasm, 9, 89
cytoprotection, 50, 51, 52, 55, 81, 115
cytotoxic agents, 52
cytotoxicity, 53, 129, 130, 138

D

disseminated intravascular coagulation, 131
DTEase enzyme (DTEase), 4, 5, 6, 7
Dubin-Johnson syndrome, 91, 113, 114

E

E. coli, 9, 11, 12, 18, 23, 26, 27, 28, 31, 35, 36, 45, 188, 197, 199
encephalopathy, xi, 90, 94, 205, 206, 211, 212, 213, 214
Enterobacterales, v, x, xi, 185, 186, 187, 188, 189, 190, 197, 199, 200, 201
entropy, 156, 157, 158, 165, 173

F

First Law of Thermodynamics, 156
fructose, 2, 3, 4, 5, 6, 7, 9, 13, 15, 17, 18, 19, 39, 41, 43, 44, 45, 46

G

Gibbs energy, 156, 160, 161, 162, 164, 168, 173, 174, 175, 177, 182
group-contribution method, 153, 157, 173, 174

H

hepatocytes, 87, 90, 91, 92, 102
hyperbilirubinemia, 90, 91, 92, 93, 95, 111
hypoxia, 50, 51, 54, 74, 76, 79, 81, 84, 96, 207, 211, 213
hypoxic-ischemic encephalopathy (HIE), xi, 205, 206, 208, 209, 210, 211, 213

I

in vitro, 73, 76, 77, 93, 110, 116, 120, 147, 188, 189, 197, 200, 201
in vivo, 51, 75, 76, 77, 93, 109, 176
infection, ix, 96, 110, 124, 126, 127, 128, 132, 134, 136, 139, 141, 143, 144, 145, 146, 201
inflammatory mediators, 130, 138, 214
inflammatory responses, 106
International Union of Biochemistry and Molecular Biology (IUBMB), 155, 160, 168, 177
International Union of Pure and Applied Chemistry (IUPAC), 155, 160, 175, 177, 178, 183
isomerization, 4, 6, 12, 13, 22, 29, 31, 33, 35, 183

J

jaundice, 87, 89, 92, 113, 115

K

K^+, vii, viii, 49, 50, 51, 53, 54, 55, 56, 58, 59, 60, 61, 62, 63, 64, 65, 66, 67, 69, 70, 71, 72, 73, 74, 76, 77, 78, 79, 80, 81, 82, 83, 169

L

leukocidin, 124, 125, 141, 142, 143, 144, 145, 146, 147, 148, 149
leukopenia, ix, 123, 129, 131, 136
linezolid, ix, 124, 134, 140, 147
L-rhamnose isomerase (L-RhI), 8, 9, 10, 11, 12, 13, 22, 30, 31, 38, 39, 41, 42, 43, 44, 46

M

metabolic syndrome, 93, 103, 119
metabolism, viii, 2, 3, 18, 20, 25, 42, 43, 74, 85, 88, 89, 90, 103, 110, 111, 112, 135, 157, 179
metabolites, 8, 154, 158, 161, 169, 174, 175
Mg^{2+}, 24, 56, 57, 58, 61, 62, 64, 65, 66, 157, 158, 168, 169, 175, 179, 181
mitochondria, viii, 49, 50, 52, 53, 54, 55, 56, 57, 58, 59, 60, 61, 62, 63, 64, 66, 67, 69, 71, 72, 73, 74, 76, 77, 78, 80, 81, 82, 83, 84, 129, 144, 207, 213, 214

N

near-physiological reference state, 158
necrosis, 51, 71, 129, 132, 206, 207, 211, 214

O

osteomyelitis, 125, 128
oxidative stress, 100, 121, 207
oxide nanoparticles, 25, 26, 44
oxygen consumption, viii, 49, 50, 60, 61, 62, 65, 66, 74, 76, 77, 82

P

Panton Valentine leukocidin (PVL), ix, 123, 124, 125, 126, 127, 128, 129, 130, 131, 132, 133, 134, 135, 136, 137, 138, 139, 140, 141, 142, 143, 146, 147
perinatal asphyxia (PA), vi, vii, xi, 101, 145, 147, 203, 205, 206, 207, 209, 212, 214
phycocyanobilin, 85, 108, 109, 110, 120, 121
piperacillin/tazobactam (PIP/TAZ), x, 185, 187, 188, 196, 197, 199, 200, 201, 204
pneumonia, vii, ix, x, 122, 123, 124, 125, 126, 127, 128, 129, 130, 131, 132, 133, 134, 135, 136, 137, 138, 139, 140, 141, 143, 144, 146, 147, 148, 149, 185, 201, 204
polymorphonuclear leukocyte (PMN), ix, 123, 129, 131, 144
potassium cycle, viii, 49, 50, 66
protein binding, 85
Pseudomonas aeruginosa (*P. aeruginosa*), xi, 186, 187, 188, 189, 195, 199, 200, 201, 202, 203

Q

quantum chemical calculations, x, 153
quinolinic acid, 214

R

reactive oxygen, 3, 41, 80, 81, 82, 83, 129, 212
reference state(s), x, 153, 154, 155, 159, 160, 163, 164, 165, 167, 168, 169, 182
resistance, 10, 38, 74, 77, 79, 84, 87, 88, 91, 117, 125, 134, 139, 142, 149, 187, 188, 200, 203, 204
respiratory distress syndrome, 208
respiratory failure, 130, 136, 137
Ribitol dehydrogenase (RDH), 13, 15, 17, 18, 19, 45

S

Second Law of Thermodynamics, 156, 157
serum albumin, viii, 85, 87, 93, 115, 116, 120
signaling pathway, 52, 53, 54, 55, 60, 71, 79, 80, 84, 114
staphylococcus aureus (*s. aureus*), 124
state, x, 7, 12, 28, 51, 54, 60, 64, 65, 67, 69, 70, 72, 73, 153, 154, 155, 156, 157, 158, 160, 162, 163, 164, 165, 167, 168, 169, 176, 179, 182
state functions, 155, 156, 157, 163
structure, vii, ix, 2, 3, 5, 6, 7, 8, 9, 11, 12, 25, 32, 34, 41, 46, 55, 59, 82, 85, 94, 96, 98, 100, 104, 107, 109, 117, 118, 119, 126, 139, 144, 186
subarachnoid hemorrhage, 79, 105, 119

T

thermal stability, 5, 7, 16, 34, 37, 44, 109
thermodynamic calculations, x, 153, 163, 164, 168
thermodynamic equilibrium, 17, 35
thermodynamic properties, 153, 154, 156, 157, 158, 159, 160, 161, 163, 164, 166, 167, 168, 169, 172, 173, 174, 175, 176, 177, 178, 179, 181, 183
thermodynamic theory, 154, 155
thermodynamics, vii, x, 153, 154, 155, 158, 163, 168, 173, 176, 177, 178, 180, 181, 182, 183
tigecycline, ix, x, 124, 134, 135, 147, 185, 188, 196, 197
toxicity, 8, 80, 89, 90, 97, 122, 129, 135
toxin, ix, 124, 126, 129, 130, 132, 134, 135, 136, 137, 138, 140, 142, 143, 145, 148
transformed thermodynamic properties, 153, 168, 173, 174, 176, 178

U

unconjugated bilirubin, viii, 85, 86, 87, 89, 90, 92, 98, 110

V

virus infection, ix, 123, 127, 128, 130, 132

X

X-ray diffraction, 46